VEDIC COSMOGRAPHY AND ASTRONOMY

VEDIC COSMOGRAPHY AND ASTRONOMY

RICHARD L. THOMPSON, PH.D.

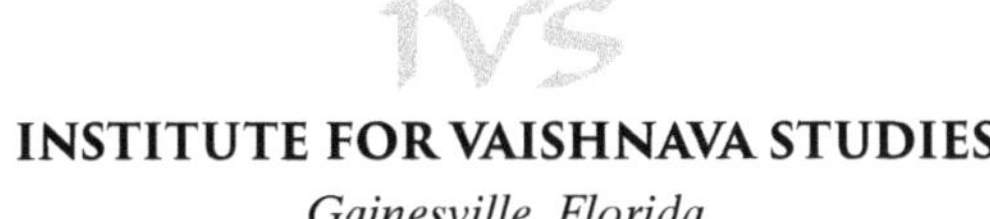

INSTITUTE FOR VAISHNAVA STUDIES
Gainesville, Florida

THE COVER: An astronomical instrument seen in Benares, India, in 1772 by an Englishman named Robert Barker. Said to be about two hundred years old at the time, the structure included two quadrants that were used to measure the position of the sun.

Figure 10, "The Armillary Sphere," is from *Astronomy*, by A. Krause, published in 1961 by Longman Group UK Ltd. Used with permission.

Figure 13, "The Norse Conception of the World," is from *Nordisch Germanische Götter*, by W. Wagner, published by Haude & Spenersche Verlagsbuchhandlung GmbH. Used with permission.

First printing 1990
Second printing 1991
Third printing 1996
Fourth printing 2017

Published by Institute for Vaishnava Studies (IVS)

Contact information:
Web: www.richardlthompson.com
Email: rlthompsonarchives@ivs.edu
Write to: Richard L. Thompson Archives,
P. O. Box 1791, Alachua, FL 32616

Library of Congress Cataloging-in-Publication Data

Thompson, Richard L.
Vedic cosmography and astronomy.
Includes bibliographical references.
1. Astronomy, Hindu. 2. Cosmography. 3. Purāṇas.
Bhāgavatapurāṇa. I. Title.
QB17.T46 1989 520'.954 89-18293
ISBN 978-0-9981871-5-0

Dedicated to

His Divine Grace
A. C. Bhaktivedanta Swami Prabhupāda

oṁ ajñāna-timirāndhasya jñānāñjana-śalākayā
cakṣur unmīlitaṁ yena tasmai śrī-gurave namaḥ

Other books by Richard L. Thompson

MECHANISTIC AND NONMECHANISTIC SCIENCE
An Investigation into the Nature of Consciousness and Form

MYSTERIES OF THE SACRED UNIVERSE
The Cosmology of the *Bhāgavata Purāṇa*

MAYA
The World as Virtual Reality

GOD AND SCIENCE
Divine Causation and the Laws of Nature

PARALLELS
Ancient Insights into Modern UFO Phenomena

FORBIDDEN ARCHEOLOGY
The Hidden History of the Human Race
(Co-authored with Michael A. Cremo)

THE HIDDEN HISTORY OF THE HUMAN RACE
(Abridged version of *Forbidden Archeology*)

CONSCIOUSNESS: THE MISSING LINK
by His Divine Grace A. C. Bhaktivedanta Swami Prabhupāda,
Dr. T. D. Singh and Richard L. Thompson

CONTENTS

INTRODUCTION

"Now our Ph.D.'s must collaborate and study the Fifth Canto to make a model for building the Vedic Planetarium. My final decision is that the universe is just like a tree, with root upwards. Just as a tree has branches and leaves, so the universe is also composed of planets which are fixed up in the tree like the leaves, flowers, fruits, etc. . . . So now all you Ph.D.'s must carefully study the details of the Fifth Canto and make a working model of the universe. If we can explain the passing seasons, eclipses, phases of the moon, passing of day and night, etc., then it will be very powerful propaganda" (letter from Śrīla Prabhupāda to Svarūpa Dāmodara dāsa, April 27, 1976).

In the year A.D. 1068 a group of workmen labored to erect an earthen mound about sixty feet high in the Anglo-Saxon village of Cambridge, northeast of London. On top of this mound they built a stone tower that dominated the small collection of thatched houses huddled alongside the river Cam. This tower served as a fortress to protect and consolidate this part of the kingdom, which William the Conqueror had won just two years before.

At this time the Western, or European, civilization, which is so important in the world today, was just beginning to emerge from the debris of previous cultures and societies. Science as we know it today was unheard of, and the Christian Church was in the process of solidifying its position in the previously pagan territories of northern Europe. The writings of the ancient Greeks and other early civilizations were largely lost, and would not be reintroduced into Europe from Arab sources for some three hundred years. Universities already existed in southern European countries; in Britain some two hundred years would pass before the founding of Oxford and then Cambridge.

In A.D. 1000, about sixty years before the erection of the stone tower on the Cam, an Arab scholar named Alberuni completed a book on India (AL). Alberuni lived in the kingdom of Ghaznia, in the court of one King Mahmud—a Muslim king who specialized in raiding the northwestern territories of India, such as Sind and the Punjab. Alberuni was a well-known scholar of his time who read Plato in the original Greek and who had also studied Sanskrit. He was apparently employed by King Mahmud to study the Hindus, in much the same way that the United States government now employs scholars to study the Russians and the Communist Chinese.

Alberuni's access to source material in Sanskrit was limited. He had access to the body of Indian astronomical literature called *jyotiṣa śāstra*, and he also had access to a number of *Purāṇas*, such as the *Matsya Purāṇa* and the *Vāyu Purāṇa*. He mentions the *Śrīmad-Bhāgavatam*, or *Bhāgavata Purāṇa*, but apparently he never saw a copy of it.

In this body of literature, Alberuni was mainly interested in information relating to the Indian view of the universe and the observable material events taking place within it. Indeed, the most striking feature of Alberuni's book is that nearly half of it is concerned with Indian astronomy and cosmology.

One important division of the *jyotiṣa śāstra* consists of works on mathematical astronomy known as astronomical *siddhāntas*. These include works of historical Indian astronomers, such as Āryabhaṭa, Brahmagupta, and Viraha Mihira, some of whom were nearly Alberuni's contemporaries. They also include ancient Sanskrit texts, such as the *Sūrya-siddhānta*, that were said to have been originally disseminated by demigods and great *ṛṣis*. These works treat the earth as a small globe floating in space and surrounded by the planets, which orbit around it. They are mainly concerned with the question of how to calculate the positions of the planets in the sky at any desired time. They contain elaborate rules for performing these calculations, as well as much numerical data concerning the distances, sizes, and rates of motion of the planets. However, they say very little about the nature of the planets, their origin, and the causes of their motion.

The calculations described in the astronomical *siddhāntas* were well understood by Alberuni, and it seems that at that time there was considerable interest in Indian astronomy in the centers of Muslim civilization. He was also familiar with the Greek astronomical tradition, epitomized by Ptolemy. However, Alberuni found the cosmology presented in the *Purāṇas* very hard to understand. His account of Purāṇic cosmology closely follows the Fifth Canto of the *Śrīmad-Bhāgavatam*, and the *Purāṇas* in general. When dealing with this material, Alberuni frequently expressed exasperation and complete incomprehension, much as many people do today, and he naturally took this as an opportunity to criticize Hindu dharma and assert the superiority of his own Muslim tradition.

In this book we will discuss the cosmology presented in the Fifth Canto of the *Śrīmad-Bhāgavatam* and try to clarify its relationship with other prominent systems of cosmology, both ancient and modern. We have begun with this historical account to show that bewilderment with the cosmology of the *Bhāgavatam* is not a new phenomenon caused by the rise of modern science. The same bewilderment also affected Alberuni, even though in his society the earth was regarded as being fixed in the center of the universe.

Many Indian astronomers of earlier centuries were also unable to understand Vedic cosmology, and they were led to openly reject parts of it, even though their own religious and social tradition was based on the *Purāṇas*. For example, Bhāskarācārya, the 11th-century author of the siddhāntic text *Siddhānta-śiromaṇi*, could not reconcile the relatively small diameter of the earth, which he deduced from simple measurements, with the immense magnitude attributed to the earth by the Paurānikas, the followers of the *Purāṇas* (SSB1, pp. 114–15). Likewise, the 15th-century south Indian astronomer Parameśvara stated that the Purāṇic account of the seven *dvīpas* and oceans is something "given only for religious meditation," and that the 84,000-*yojana* height of Mount Meru described in the *Purāṇas* is "not acceptable to the astronomers" (GP, pp. 85, 87).

Vaiṣṇavas of past centuries also discussed the relationship between the Fifth Canto of *Śrīmad-Bhāgavatam* and the *jyotiṣa śāstras*. An example of this is found in the *Bhāgavatam* commentary of Vaṁśīdhara, a Vaiṣṇava who lived in the 17th century A.D. In this commentary, Vaṁśīdhara discusses the apparent conflict between the small size of the earth, as described in the *jyotiṣa śāstras*, and the large size of Bhū-maṇḍala, as described in the Fifth Canto. His analysis of this apparent conflict is discussed in Appendix 1.

There are evidently serious disagreements between the cosmological system of the *Purāṇas* and the world models that human observers tend to arrive at using their reasoning powers and their ordinary senses. The cause of these difficulties is not simply the rise of modern Western science. They have existed in India since a time antedating the rise of modern Western culture, and to some they may seem to be based on an inherent contradiction within the Vedic tradition itself.

The long-standing perplexity that has attended the subject of Vedic cosmology indicates that these disagreements are very deep and difficult to resolve. However, the thesis of this book is that the disagreements are not irreconcilable. The apparent contradictions can be resolved by developing a proper understanding of the nature of space, time, and matter, as described in the *Śrīmad-Bhāgavatam*, and a corresponding understanding of the Vedic approach to describing and thinking about reality.

In Chapter 1 we begin our account of Vedic astronomy by discussing the astronomical *siddhāntas*. We give evidence indicating that these works form an integral part of the original Vedic tradition. To accept these works and reject Purāṇic cosmology, as some Indian astronomers have done, is to start down the path of modern scientific materialism, which ultimately leads to the total rejection of the Vedic literature. But to reject the astronomical *siddhāntas* as anti-Vedic means to lose the Vedic tradition of rigorous mathematical astronomy. This plays into the hands of the modern Western scholars who wish to reject the *Vedas* and *Purāṇas* as mythological, and who interpret the astronomical *siddhāntas* as products of Greek scientific genius that were borrowed and falsely dressed in Hindu garb by dishonest *brāhmaṇas*. (In Appendix 2 we address some of the arguments of these scholars and show that they are seriously flawed.)

Our thesis is that the astronomical *siddhāntas* and the Purāṇic cosmology can be understood as mutually compatible accounts of one multifaceted material reality. Modern Western science is based on the idea that nature can be fully described by a single, rational world-model. However, the *Śrīmad-Bhāgavatam* points out that no person of this world is capable of fully describing the material universe "even in a lifetime as long as that of Brahmā" (SB 5.16.4). Thus the Vedic approach to the description of nature is based on the strategy of presenting many mutually compatible aspects of one humanly indescribable complete whole.

The old story of the blind men and the elephant epitomizes this approach. Each blind man observed a genuine aspect of the elephant, and a seeing man could understand how all of these aspects fit together to form a coherent whole. Even a blind man, after carefully studying the reports coming from the seeing man and his fellow blind men, could begin to understand the nature of the whole elephant, although he could not directly sense it without obtaining a cure for his blindness. We suggest that in our attempts to understand the material universe, we are comparable to a blind man feeling a particular part of the elephant.

According to this analogy, the astronomical *siddhāntas* present the cosmos as it appears to similar blind men of this earth, and literatures such as the *Bhāgavatam* present the world view of beings with higher powers of vision. These include demigods, *ṛṣis*, and ultimately the Supreme Lord, who alone can see the entire universe. These higher beings can directly see both the aspects of the universe presented in the *Bhāgavatam* and the aspects presented in the astronomical *siddhāntas*. To these higher beings it is apparent how all of these aspects fit together consistently in a complete whole, even though we can begin to understand this only with great effort.

We note that with the development of modern physics, scientists have at least temporarily been forced to abandon the goal of formulating one complete mathematical model of the atom. According to the standard interpretation of the quantum theory introduced by Niels Bohr, atomic phenomena must be understood from at least two complementary perspectives rather than as a single, intelligible whole. These perspectives—the wave picture and the particle picture—seem to contradict each other, and yet they are both valid descriptions of nature. They are facets of a coherent theory of the atom, but they cannot be combined within the framework of classical physics. To unite them and show their compatibility, one must go to a higher-dimensional level of mathematical abstraction, which is very difficult to comprehend.

In developing an understanding of Vedic cosmology as a multifaceted description of reality, it will be necessary to free ourselves from the rigid framework of Cartesian and Euclidian three-dimensional geometry, which forms the basis of the modern scientific world view. We will attempt to do this in Chapter 2, where we will discuss space, physical laws, and processes of sense perception, as presented in the *Śrīmad-Bhāgavatam*. In Chapters 3 and 4 we will give an account of Purāṇic cosmology and show how the ideas developed in Chapter 2 can be applied to resolve apparent contradictions within the Vedic tradition and between the Vedic cosmology and the world of our ordinary sensory experience. Here a key idea is that the universe as described in Vedic literature is higher-dimensional: it cannot be fully represented within three-dimensional space.

In our discussion of Vedic cosmology we will be forced to interpret the texts of the *Śrīmad-Bhāgavatam* and other Vedic literature. This is inevitable, since even a literal interpretation is based on underlying assumptions made by the reader—assumptions that may differ from those of the author of the text, and that the reader may hold without being consciously aware of them. In making such interpretations we will try to adhere to the following rule given by Śrīla Prabhupāda: "*The original purpose of the text must be maintained.* No obscure meaning should be screwed out of it, yet it should be presented in an interesting manner for the understanding of the audience. This is called realization" (SB 1.4.1p). We also note that Śrīla Prabhupāda advocated in SB 5.16.10p that we should accept the cosmological statements in the *Śrīmad-Bhāgavatam* as authoritative and simply try to appreciate them. We will therefore adopt the working assumption that even though these statements may seem very hard to comprehend, they nonetheless do present an understandable and realistic description of the universe.

In Chapter 5 we address the question of whether or not there is any empirical evidence supporting the higher-dimensional picture of the universe that we derive from the *Śrīmad-Bhāgavatam*. It turns out that there is voluminous evidence along these lines, although practically none of it is accepted by the scientific community.

In Chapter 6 we return to Vedic cosmology and discuss a number of controversial topics, including gravitation, the moon flight, the scale of cosmic distances, and the nature of stars. In Chapter 7 we survey the modern scientific evidence regarding the theory of the expanding universe. Here we not only find that this theory is flawed, but we also find evidence indicating that Newton's laws of motion fail on the galactic level. Finally, in Chapter 8 we present brief answers to a number of common questions.

The material presented in this book constitutes a preliminary study of Vedic cosmology and astronomy. To properly answer the many questions that arise, much further research will have to be done. This will include (1) careful study of cosmological material in a wide variety of Vedic literatures, (2) study of Vedic geographical material, (3) careful analysis of the theories of Western scholars about the history of Vedic astronomy, (4) study of ancient astronomical observations, (5) study of dating and the Vedic calendar, (6) study of empirical evidence relating to Vedic cosmology, and (7) the careful analysis of modern cosmology and astronomy. It is our hope that these studies will culminate in the development of a Vedic planetarium and museum that can effectively present Kṛṣṇa consciousness in the context of Vedic cosmology. This, of course, was Śrīla Prabhupāda's plan for the planetarium in the Temple of Understanding in Śrīdhāma Māyāpura, and similar planetariums can be set up in cities around the world.

In this book we will use the terms *Vedic* and *Purāṇic* interchangeably. Although modern scholars reject this usage, it is justified by the verse *itihāsa-purāṇaṁ ca pañcamo veda ucyate* in *Śrīmad-Bhāgavatam* (1.4.20). According to this verse, the *Purāṇas* and the histories, such as the *Mahābhārata*, are known as the fifth *Veda*. References to Sanskrit and Bengali texts are of three forms: A reference such as SB 5.22.14 means that the quotation is from the 14th verse of Chapter 22 of the Fifth Canto of *Śrīmad-Bhāgavatam*. A reference such as SB 5.21.6p means the quotation is from Śrīla Prabhupāda's purport to verse 6 of Chapter 21 of the Fifth Canto. And a reference such as SB 5.21cs means the quotation is from the Chapter Summary of Chapter 21 of the Fifth Canto. AL or ML after references to the *Caitanya-caritāmṛta* indicate *Ādi-līlā* or *Madhya-līlā*. For books not divided into verses and purports, we cite the code identifying the book, followed by the page number (see the Bibliography).

The Astronomical Siddhāntas

Since the cosmology of the astronomical *siddhāntas* is quite similar to traditional Western cosmology, we will begin our discussion of Vedic astronomy by briefly describing the contents of these works and their status in the Vaiṣṇava tradition. In a number of purports in the *Caitanya-caritāmṛta*, Śrīla Prabhupāda refers to two of the principal works of this school of astronomy, the *Sūrya-siddhānta* and the *Siddhānta-śiromaṇi*. The most important of these references is the following:

> These calculations are given in the authentic astronomy book known as the *Sūrya-siddhānta*. This book was compiled by the great professor of astronomy and mathematics Bimal Prasād Datta, later known as Bhaktisiddhānta Sarasvatī Gosvāmī, who was our merciful spiritual master. He was honored with the title Siddhānta Sarasvatī for writing the *Sūrya-siddhānta*, and the title Gosvāmi Mahārāja was added when he accepted *sannyāsa*, the renounced order of life [CC AL 3.8p].

Here the *Sūrya-siddhānta* is clearly endorsed as an authentic astronomical treatise, and it is associated with Śrīla Bhaktisiddhānta Sarasvatī Ṭhākura. The *Sūrya-siddhānta* is an ancient Sanskrit work that, according to the text itself, was spoken by a messenger from the sun-god, Sūrya, to the famous *asura* Maya Dānava at the end of the last Satya-yuga. It was translated into Bengali by Śrīla Bhaktisiddhānta Sarasvatī, who was expert in Vedic astronomy and astrology.

Some insight into Śrīla Bhaktisiddhānta's connection with Vedic astronomy can be found in the bibliography of his writings. There it is stated,

> In 1897 he opened a "Tol" named "Saraswata Chatuspati" in Manicktola Street for teaching Hindu Astronomy nicely calculated independently of Greek and other European astronomical findings and calculations.
>
> During this time he used to edit two monthly magazines named "Jyotirvid" and "Brihaspati" (1896), and he published several authoritative treatises on Hindu Astronomy. . . . He was offered a chair in the Calcutta University by Sir Asutosh Mukherjee, which he refused [BS1, pp. 2–3].

These statements indicate that Śrīla Bhaktisiddhānta took considerable interest in Vedic astronomy and astrology during the latter part of the nineteenth century, and they suggest that one of his motives for doing this was to establish that the Vedic astronomical tradition is independent of Greek and European influence. In addition to his Bengali translation of the *Sūrya-siddhānta*, Śrīla Bhaktisiddhānta Sarasvatī published the following works in his two magazines:

> (a) Bengali translation and explanation of Bhāskarācārya's *Siddhānta-Shiromani Goladhyaya* with *Basanabhasya*, (b) Bengali translation of *Ravichandrasayanaspashta, Laghujatak*, with annotation of Bhattotpala, (c) Bengali translation of *Laghuparashariya*, or *Ududaya-Pradip*, with Bhairava Datta's annotation, (d) Whole of *Bhauma-Siddhānta* according to western calculation, (e) Whole of *Ārya-Siddhānta* by Āryabhaṭa, (f) Paramadishwara's *Bhatta Dipika-Tika, Dinakaumudi, Chamatkara-Chintamoni*, and *Jyotish-Tatwa-Samhita* [BS1, p. 26].

This list includes a translation of the *Siddhānta-śiromaṇi*, by the 11th-century astronomer Bhāskarācārya, and the *Ārya-siddhānta*, by the 6th-century astronomer Āryabhaṭa. Bhaṭṭotpala was a well-known astronomical commentator who lived in the 10th century. The other items in this list also deal with astronomy and astrology, but we do not have more information regarding them.

Śrīla Bhaktisiddhānta Sarasvatī also published the *Bhaktibhāvana Pañjikā* and the *Śrī Navadvīpa Pañjikā* (BS2, pp. 56, 180). A *pañjikā* is an almanac that includes dates for religious festivals and special days such as Ekādaśī. These dates are traditionally calculated using the rules given in the *jyotiṣa śāstras*.

During the time of his active preaching as head of the Gauḍīya Math, Śrīla Bhaktisiddhānta stopped publishing works dealing specifically with astronomy and astrology. However, as we will note later on, Śrīla Bhaktisiddhānta cites both the *Sūrya-siddhānta* and the *Siddhānta-śiromaṇi* several times in his *Anubhāṣya* commentary on the *Caitanya-caritāmṛta*.

It is clear that in recent centuries the *Sūrya-siddhānta* and similar works have played an important role in Indian culture. They have been regularly used for preparing calendars and for performing astrological calculations. In Section 1.c we cite evidence from the *Bhāgavatam* suggesting that complex astrological and calendrical calculations were also regularly performed in Vedic times. We therefore suggest that similar or identical systems of astronomical calculation must have been known in this period.

Here we should discuss a potential misunderstanding. We have stated that Vaiṣṇavas have traditionally made use of the astronomical *siddhāntas* and that both Śrīla Prabhupāda and Śrīla Bhaktisiddhānta Sarasvatī Ṭhākura have referred to them. At the same time, we have pointed out that the authors of the astronomical *siddhāntas*, such as Bhāskarācārya, have been unable to accept some of the cosmological statements in the *Purāṇas*. How could Vaiṣṇava acharyas accept works which criticize the *Purāṇas*?

We suggest that the astronomical *siddhāntas* have a different status than transcendental literature such as the *Śrīmad-Bhāgavatam*. They are authentic in the sense that they belong to a genuine Vedic astronomical tradition, but they are nonetheless human works that may contain imperfections. Many of these works, such as the *Siddhānta-śiromaṇi*, were composed in recent centuries and make use of empirical observations. Others, such as the *Sūrya-siddhānta*, are attributed to demigods but were transmitted to us by persons who are not spiritually perfect. Thus the *Sūrya-siddhānta* was recorded by Maya Dānava. Śrīla Prabhupāda has said that Maya Dānava "is always materially happy because he is favored by Lord Śiva, but he cannot achieve spiritual happiness at any time" (SB 5.24cs).

The astronomical *siddhāntas* constitute a practical division of Vedic science, and they have been used as such by Vaiṣṇavas throughout history. The thesis of this book is that these works are surviving remnants of an earlier astronomical science that was fully compatible with the cosmology of the *Purāṇas*, and that was disseminated in human society by demigods and great sages. With the progress of Kali-yuga, this astronomical knowledge was largely lost. In recent centuries the knowledge that survived was reworked by various Indian astronomers and brought up to date by means of empirical observations.

Although we do not know anything about the methods of calculation used before the Kali-yuga, they must have had at least the same scope and order of sophistication as the methods presented in the *Sūrya-siddhānta*. Otherwise they could not have produced comparable results. In presently available Vedic literature, such computational methods are presented only in the astronomical *siddhāntas* and other *jyotiṣa śāstras*. The *Itihāsas* and *Purāṇas* (including the *Bhāgavatam*) do not contain rules for astronomical calculations, and the *Vedas* contain only the *Vedāṅga-jyotiṣa*, which is a *jyotiṣa śāstra* but is very brief and rudimentary (VJ).

The following is a brief summary of the topics covered by the *Sūrya-siddhānta:* (1) computation of the mean and true positions of the planets in the sky, (2) determination of latitude and longitude and local celestial coordinates, (3) prediction of full and partial eclipses of the moon and sun, (4) prediction of conjunctions of planets with stars and other planets, (5) calculation of the rising and setting times of planets and stars, (6) calculation of the moon's phases, (7) calculation of the dates of various astrologically significant planetary combinations (such as Vyatīpāta), (8) a discussion of cosmography, (9) a discussion of astronomical instruments, and (10) a discussion of kinds of time.

We will first discuss the computation of mean and true planetary positions, since it introduces the *Sūrya-siddhānta's* basic model of the planets and their motion in space.

1.A. THE SOLAR SYSTEM ACCORDING TO THE SŪRYA–SIDDHĀNTA

The *Sūrya-siddhānta* treats the earth as a globe fixed in space, and it describes the seven traditional planets (the sun, the moon, Mars, Mercury, Jupiter, Venus, and Saturn) as moving in orbits around the earth. It also describes the orbit of the planet Rāhu, but it makes no mention of Uranus, Neptune, and Pluto. The main function of the *Sūrya-siddhānta* is to provide rules allowing us to calculate the positions of these planets at any given time. Given a particular date, expressed in days, hours, and minutes since the beginning of Kali-yuga, one can use these rules to compute the direction in the sky in which each of the seven planets will be found at that time. All of the other calculations described above are based on these fundamental rules.

The basis for these rules of calculation is a quantitative model of how the planets move in space. This model is very similar to the modern Western model of the solar system. In fact, the only major difference between these two models is that the *Sūrya-siddhānta's* is geocentric, whereas the model of the solar system that forms the basis of modern astronomy is heliocentric.

To determine the motion of a planet such as Venus using the modern heliocentric system, one must consider two motions: the motion of Venus around the sun and the motion of the earth around the sun. As a crude first approximation, we can take both of these motions to be circular. We can also imagine that the earth is stationary and that Venus is revolving around the sun, which in turn is revolving around the earth. The relative motions of the earth and Venus are the same, whether we adopt the heliocentric or geocentric point of view.

In the *Sūrya-siddhānta* the motion of Venus is also described, to a first approximation, by a combination of two motions, which we can call cycles 1 and 2. The first motion is in a circle around the earth, and the second is in a circle around a point on the circumference of the first circle. This second circular motion is called an epicycle.

It so happens that the period of revolution for cycle 1 is one earth year, and the period for cycle 2 is one Venusian year, or the time required for Venus to orbit the sun according to the heliocentric model. Also, the sun is located at the point on the circumference of cycle 1 which serves as the center of rotation for cycle 2. Thus we can interpret the *Sūrya-siddhānta* as saying that Venus is revolving around the sun, which in turn is revolving around the earth (see Figure 1). According to this interpretation, the only difference between the

TABLE 1

Planetary Years, Distances, and Diameters, According to Modern Western Astronomy

Planet	Length of Year	Mean Distance from Sun	Mean Distance from Earth	Diameter
Sun	—	0.	1.00	865,110
Mercury	87.969	.39	1.00	3,100
Venus	224.701	.72	1.00	7,560
Earth	365.257	1.00	0.	7,928
Mars	686.980	1.52	1.52	4,191
Jupiter	4,332.587	5.20	5.20	86,850
Saturn	10,759.202	9.55	9.55	72,000
Uranus	30,685.206	19.2	19.2	30,000
Neptune	60,189.522	30.1	30.1	28,000
Pluto	90,465.38	39.5	39.5	?

Years are equal to the number of earth days required for the planet to revolve once around the sun. Distances are given in astronomical units (AU), and 1 AU is equal to 92.9 million miles, the mean distance from the earth to the sun. Diameters are given in miles. (The years are taken from the standard astronomy text TSA, and the other figures are taken from EA.)

Sūrya-siddhānta model and the modern heliocentric model is one of relative point of view.

In Tables 1 and 2 we list some modern Western data concerning the sun, the moon, and the planets, and in Table 3 we list some data on periods of planetary revolution taken from the *Sūrya-siddhānta*. The periods for cycles 1 and 2 are given in revolutions per *divya-yuga*. One *divya-yuga* is 4,320,000 solar years, and a solar year is the time it takes the sun to make one complete circuit through the sky against the background of stars. This is the same as the time it takes the earth to complete one orbit of the sun according to the heliocentric model.

For Venus and Mercury, cycle 1 corresponds to the revolution of the earth around the sun, and cycle 2 corresponds to the revolution of the planet around the sun. The times for cycle 1 should therefore be one revolution per solar year, and, indeed, they are listed as 4,320,000 revolutions per *divya-yuga*.

The times for cycle 2 of Venus and Mercury should equal the modern heliocentric years of these planets. According to the *Sūrya-siddhānta*, there are 1,577,917,828 solar days per *divya-yuga*. (A solar day is the time from sunrise to sunrise.) The cycle-2 times can be computed in solar days by dividing this

TABLE 2

Data Pertaining to the Moon, According to Modern Western Astronomy

Siderial Period	27.32166 days
Synodic Period	29.53059 days
Nodal Period	27.2122 days
Siderial Period of Nodes	−6,792.28 days
Mean Distance from Earth	238,000 miles = .002567 AU
Diameter	2,160 miles

The sidereal period is the time required for the moon to complete one orbit against the background of stars. The synodic period, or month, is the time from new moon to new moon. The nodal period is the time required for the moon to pass from ascending node back to ascending node. The sidereal period of the nodes is the time for the ascending node to make one revolution with respect to the background of stars. (This is negative since the motion of the nodes is retrograde.) (EA)

number by the revolutions per *divya-yuga* in cycle 2. The cycle-2 times are listed as "SS [*Sūrya-siddhānta*] Period," and they are indeed very close to the heliocentric years, which are listed as "W [Western] Period" in Table 3.

For Mars, Jupiter, and Saturn, cycle 1 corresponds to the revolution of the planet around the sun, and cycle 2 corresponds to the revolution of the earth around the sun. Thus we see that cycle 2 for these planets is one solar year (or 4,320,000 revolutions per *divya-yuga*). The times for cycle 1 in solar days can also be computed by dividing the revolutions per *divya-yuga* of cycle 1 into 1,577,917,828, and they are listed under "SS Period." We can again see that they are very close to the corresponding heliocentric years.

For the sun and moon, cycle 2 is not specified. But if we divide 1,577,917,828 by the numbers of revolutions per *divya-yuga* for cycle 1 of the sun and moon, we can calculate the number of solar days in the orbital periods of these planets. Table 3 shows that these figures agree well with the modern values, especially in the case of the moon. (Of course, the orbital period of the sun is simply one solar year.)

In Table 3 a cycle-1 value is also listed for the planet Rāhu. Rāhu is not recognized by modern Western astronomers, but its position in space, as described in the *Sūrya-siddhānta*, does correspond with a quantity that is measured by modern astronomers. This is the ascending node of the moon.

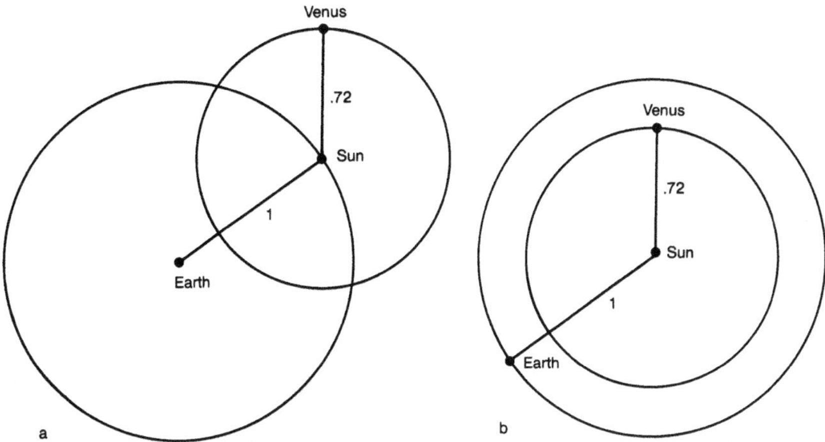

Figure 1. The geocentric and heliocentric models of the motion of Venus: a) the geocentric model of the *Sūrya-siddhānta*; b) the heliocentric model.

TABLE 3

Planetary Periods According to the Sūrya-siddhānta

Planet	Cycle 1	Cycle 2	SS Period	W Period
Moon	57,753,336	*	27.322	27.32166
Mercury	4,320,000	17,937,000	87.97	87.969
Venus	4,320,000	7,022,376	224.7	224.701
Sun	4,320,000	*	365.26	365.257
Mars	2,296,832	4,320,000	687.0	686.980
Jupiter	364,220	4,320,000	4,332.3	4,332.587
Saturn	146,568	4,320,000	10,765.77	10,759.202
Rāhu	–232,238	*	–6,794.40	–6,792.280

The figures for cycles 1 and 2 are in revolutions per *divya-yuga*. The "SS Period" is equal to 1,577,917,828, the number of solar days in a *yuga* cycle, divided by one of the two cycle figures (see the text). This should give the heliocentric period for Mercury, Venus, the earth (under sun) Mars, Jupiter, and Saturn, and it should give the geocentric period for the moon and Rāhu. These periods can be compared with the years in Table 1 and the sidereal periods of the moon and its nodes in Table 2. These quantities have been reproduced from Tables 1 and 2 in the column labeled "W Period."

From a geocentric perspective, the orbit of the sun defines one plane passing through the center of the earth, and the orbit of the moon defines another such plane. These two planes are slightly tilted with respect to each other, and thus they intersect on a line. The point where the moon crosses this line going from celestial south to celestial north is called the ascending node of the moon. According to the *Sūrya-siddhānta*, the planet Rāhu is located in the direction of the moon's ascending node.

From Table 3 we can see that the modern figure for the time of one revolution of the moon's ascending node agrees quite well with the time for one revolution of Rāhu. (These times have minus signs because Rāhu orbits in a direction opposite to that of all the other planets.)

TABLE 4

Heliocentric Distances of Planets, According to the Sūrya-siddhānta

Planet	Cycle 1	Cycle 2	SS Distance	W Distance
Mercury	360	133 132	.368	.39
Venus	360	262 260	.725	.72
Mars	360	235 232	1.54	1.52
Jupiter	360	70 72	5.07	5.20
Saturn	360	39 40	9.11	9.55

These are the distances of the planets from the sun. The mean heliocentric distance of Mercury and Venus in AU should be given by its mean cycle-2 circumference divided by its cycle-1 circumference. (The cycle-2 circumferences vary between the indicated limits, and we use their average values.) For the other planets the mean heliocentric distance should be the reciprocal of this (see the text). These figures are listed as "SS Distance," and the corresponding modern Western heliocentric distances are listed under "W Distance."

If cycle 1 for Venus corresponds to the motion of the sun around the earth (or of the earth around the sun), and cycle 2 corresponds to the motion of Venus around the sun, then we should have the following equation:

$$\frac{\text{circumference of cycle 2}}{\text{circumference of cycle 1}} = \frac{\text{Venus-to-Sun distance}}{\text{Earth-to-Sun distance}}$$

Here the ratio of distances equals the ratio of circumferences, since the circumference of a circle is 2π times its radius. The ratio of distances is equal to the distance from Venus to the sun in astronomical units (AU), or units of the earth-sun distance. The modern values for the distances of the planets from the sun are listed in Table 1. In Table 4, the ratios on the left of our equation are computed for Mercury and Venus, and we can see that they do agree well with the modern distance figures. For Mars, Jupiter, and Saturn, cycles 1 and 2 are switched, and thus we are interested in comparing the heliocentric distances with the reciprocal of the ratio on the left of the equation. These quantities are listed in the table, and they also agree well with the modern values. Thus, we can conclude that the *Sūrya-siddhānta* presents a picture of the relative motions and positions of the planets Mercury, Venus, Earth, Mars, Jupiter, and Saturn that agrees quite well with modern astronomy.

1.B. THE OPINION OF WESTERN SCHOLARS

This agreement between Vedic and Western astronomy will seem surprising to anyone who is familiar with the cosmology described in the Fifth Canto of the *Śrīmad-Bhāgavatam* and in the other *Purāṇas*, the *Mahābhārata*, and the *Rāmāyaṇa*. The astronomical *siddhāntas* seem to have much more in common with Western astronomy than they do with Purāṇic cosmology, and they seem to be even more closely related with the astronomy of the Alexandrian Greeks. Indeed, in the opinion of modern Western scholars, the astronomical school of the *siddhāntas* was imported into India from Greek sources in the early centuries of the Christian era. Since the *siddhāntas* themselves do not acknowledge this, these scholars claim that Indian astronomers, acting out of chauvinism and religious sentiment, Hinduized their borrowed Greek knowledge and claimed it as their own. According to this idea, the cosmology of the *Purāṇas* represents an earlier, indigenous phase in the development of Hindu thought, which is entirely mythological and unscientific.

This, of course, is not the traditional Vaiṣṇava viewpoint. The traditional viewpoint is indicated by our observations regarding the astronomical studies of Śrīla Bhaktisiddhānta Sarasvatī Ṭhākura, who founded a school for "teaching Hindu Astronomy nicely calculated independently of Greek and other European astronomical findings and calculations."

The *Bhāgavatam* commentary of the Vaiṣṇava scholar Vaṁśīdhara also sheds light on the traditional understanding of the *jyotiṣa śāstras*. His commentary appears in the book of *Bhāgavatam* commentaries Śrīla Prabhupāda used when writing his purports. In Appendix 1 we discuss in detail Vaṁśīdhara's commentary on SB 5.20.38. Here we note that Vaṁśīdhara declares the *jyotiṣa śāstra* to be the "eye of the *Vedas*," in accord with verse 1.4 of the *Nārada-saṁhitā*, which says, "The excellent science of astronomy comprising *siddhānta*, *saṁhitā*, and *horā* as its three branches is the clear eye of the *Vedas*" (BJS, xxvi).

Vaiṣṇava tradition indicates that the *jyotiṣa śāstra* is indigenous to Vedic culture, and this is supported by the fact that the astronomical *siddhāntas* do not acknowledge foreign source material. The modern scholarly view that all important aspects of Indian astronomy were transmitted to India from Greek sources is therefore tantamount to an accusation of fraud. Although scholars of the present day do not generally declare this openly in their published writings, they do declare it by implication, and the accusation was explicitly made by the first British Indologists in the early nineteenth century.

John Bentley was one of these early Indologists, and it has been said of his work that "he thoroughly misapprehended the character of the Hindu astronomical literature, thinking it to be in the main a mass of forgeries framed for the purpose of deceiving the world respecting the antiquity of the Hindu people" (HA, p. 3). Yet the modern scholarly opinion that the *Bhāgavatam* was written after the ninth century A.D. is tantamount to accusing it of being a similar forgery. In fact, we would suggest that the scholarly assessment of Vedic astronomy is part of a general effort on the part of Western scholars to dismiss the Vedic literature as a fraud.

A large book would be needed to properly evaluate all of the claims made by scholars concerning the origins of Indian astronomy. In Appendix 2 we indicate the nature of many of these claims by analyzing three cases in detail. Our observation is that scholarly studies of Indian astronomy tend to be based on imaginary historical reconstructions that fill the void left by an almost total lack of solid historical evidence.

Here we will simply make a few brief observations indicating an alternative to the current scholarly view. We suggest that the similarity between the *Sūrya-siddhānta* and the astronomical system of Ptolemy is not due to a one-sided transfer of knowledge from Greece and Alexandrian Egypt to India. Due partly to the great social upheavals following the fall of the Roman Empire, our knowledge of ancient Greek history is extremely fragmentary. However, although history books do not generally acknowledge it, evidence does exist of extensive contact between India and ancient Greece. (For example, see PA, where it is suggested that Pythagoras was a student of Indian philosophy and that *brāhmaṇas* and yogis were active in the ancient Mediterranean world.)

We therefore propose the following tentative scenario for the relations between ancient India and ancient Greece: SB 1.12.24p says that the Vedic king Yayāti was the ancestor of the Greeks, and SB 2.4.18p says that the Greeks were once classified among the *kṣatriya* kings of Bhārata but later gave up brahminical culture and became known as *mlecchas*. We therefore propose that the Greeks and the people of India once shared a common culture, which included knowledge of astronomy. Over the course of time, great cultural divergences developed, but many common cultural features remained as a result of shared ancestry and later communication. Due to the vicissitudes of the Kali-yuga, astronomical knowledge may have been lost several times in Greece over the last few thousand years and later regained through communication with India, discovery of old texts, and individual creativity. This brings us down to the late Roman period, in which Greece and India shared similar astronomical systems. The scenario ends with the fall of Rome, the burning of the famous library at Alexandria, and the general destruction of records of the ancient past.

According to this scenario, much creative astronomical work was done by Greek astronomers such as Hipparchus and Ptolemy. However, the origin of many of their ideas is simply unknown, due to a lack of historical records. Many of these ideas may have come from indigenous Vedic astronomy, and many may also have been developed independently in India and the West. Thus we propose that genuine traditions of astronomy existed both in India and the eastern Mediterranean, and that charges of wholesale unacknowledged cultural borrowing are unwarranted.

1.C. THE VEDIC CALENDAR AND ASTROLOGY

In this subsection we will present some evidence from Śrīla Prabhupāda's books suggesting that astronomical computations of the kind presented in the astronomical *siddhāntas* were used in Vedic times. As we have pointed out, many of the existing astronomical *siddhāntas* were written by recent Indian astronomers. But if the Vedic culture indeed dates back thousands of years, as the *Śrīmad-Bhāgavatam* describes, then this evidence suggests that methods of astronomical calculation as sophisticated as those of the astronomical *siddhāntas* were also in use in India thousands of years ago. Consider the following passage from the *Śrīmad-Bhāgavatam*:

> One should perform the *śrāddha* ceremony on the Makara-saṅkrānti or on the Karkaṭa-saṅkrānti. One should also perform this ceremony on the Meṣa-saṅkrānti day and the Tulā-saṅkrānti day, in the yoga named Vyatīpāta, on that day in which three lunar *tithis* are conjoined, during an eclipse of either the moon or the sun, on the twelfth lunar day, and in the Śravaṇa-nakṣatra. One should perform this ceremony on the Akṣaya-tṛtīyā day, on the ninth lunar day of the bright fortnight of the month of Kārtika, on the four *aṣṭakās* in the winter season and cool season, on the seventh lunar day of the bright fortnight of the month of Māgha, during the conjunction of Māgha-nakṣatra and the full-moon day, and on the days when the moon is completely full, or not quite completely full, when these days are conjoined with the *nakṣatras* from which the names of certain months are derived. One should also perform the *śrāddha* ceremony on the twelfth lunar day when it is in conjunction with any of the *nakṣatras* named Anurādhā, Śravaṇa, Uttara-phalgunī, Uttarāṣādhā, or Uttara-bhādrapadā. Again, one should

> perform this ceremony when the eleventh lunar day is in conjunction with either Uttara-phalgunī, Uttarāṣādhā, or Uttara-bhādrapadā. Finally, one should perform this ceremony on days conjoined with one's own birth star [*janma-nakṣatra*] or with Śravaṇa-nakṣatra [SB 7.14.20–23].

This passage indicates that to observe the *śrāddha* ceremony properly one would need the services of an expert astronomer. The *Sūrya-siddhānta* contains rules for performing astronomical calculations of the kind required here, and it is hard to see how these calculations could be performed without some computational system of equal complexity. For example, in the *Sūrya-siddhānta* the Vyatīpāta yoga is defined as the time when "the moon and sun are in different *ayanas*, the sum of their longitudes is equal to 6 signs (nearly) and their declinations are equal" (SS, p. 72). One could not even define such a combination of planetary positions without considerable astronomical sophistication.

Similar references to detailed astronomical knowledge are scattered throughout the *Bhāgavatam*. For example, the Vyatīpāta yoga is also mentioned in SB 4.12.49–50. And KB, p. 693 describes that in Kṛṣṇa's time, people from all over India once gathered at Kurukṣetra on the occasion of a total solar eclipse that had been predicted by astronomical calculation. Also, SB 10.28.7p recounts how Nanda Mahārāja once bathed too early in the Yamunā River—and was thus arrested by an agent of Varuṇa—because the lunar day of Ekādaśī ended at an unusually early hour on that occasion. We hardly ever think of astronomy in our modern day-to-day lives, but it would seem that in Vedic times daily life was constantly regulated in accordance with astronomical considerations.

The role of astrology in Vedic culture provides another line of evidence for the existence of highly developed systems of astronomical calculation in Vedic times. The astronomical *siddhāntas* have been traditionally used in India for astrological calculations, and astrology in its traditional form would be impossible without the aid of highly accurate systems of astronomical computation. Śrīla Prabhupāda has indicated that astrology played an integral role in the *karma-kāṇḍa* functions of Vedic society. A few references indicating the importance of astrology in Vedic society are SB 1.12.12p, 1.12.29p, 1.19.9–10p, 6.2.26p, 9.18.23p, 9.20.37p, and 10.8.5, and also CC AL 13.89–90 and 17.104.

These passages indicate that the traditions of the Vaiṣṇavas are closely tied in with the astronomical *siddhāntas*. Western scholars will claim that this close association is a product of processes of "Hindu syncretism" that occurred well within the Christian era and were carried out by unscrupulous *brāhmaṇas* who misappropriated Greek astronomical science and also concocted scriptures such as the *Śrīmad-Bhāgavatam*. However, if the Vaiṣṇava tradition is indeed genuine, then this association must be real, and must date back for many thousands of years.

1.D. THE STARTING DATE OF KALI–YUGA

Imagine the following scene: It is midnight on the meridian of Ujjain in India on February 18, 3102 B.C. The seven planets, including the sun and moon, cannot be seen since they are all lined up in one direction on the other side of the earth. Directly overhead the dark planet Rāhu hovers invisibly in the blackness of night.

According to the *jyotiṣa śāstras*, this alignment of the planets actually occurred on this date, which marks the beginning of the Kali-yuga. In fact, in the *Sūrya-siddhānta*, time is measured in days since the start of Kali-yuga, and it is assumed that the positions of the seven planets in their two cycles are all aligned with the star Zeta Piscium at day zero. This star, which is known as Revatī in Sanskrit, is used as the zero point for measuring celestial longitudes in the *jyotiṣa śāstras*. The position of Rāhu at day zero is also assumed to be 180 degrees from this star. Nearly identical assumptions are made in other astronomical *siddhāntas*. (In some systems, such as that of Āryabhaṭa, it is assumed that Kali-yuga began at sunrise rather than at midnight. In others, a close alignment of the planets is assumed at this time, rather than an exact alignment.)

In the *Caitanya-caritāmṛta* AL 3.9–10, the present date in this day of Brahmā is defined as follows: (1) The present Manu, Vaivasvata, is the seventh, (2) 27 *divya-yugas* of his age have passed, and (3) we are in the Kali-yuga of the 28th *divya-yuga*. The *Sūrya-siddhānta* also contains this information, and its calculations of planetary positions require knowledge of the *ahargana*, or the exact number of elapsed days in Kali-yuga. The Indian astronomer Āryabhaṭa wrote that he was 23 years old when 3,600 years of Kali-yuga had passed (BJS, part 2, p. 55). Since Āryabhaṭa is said to have been born in Śaka 398, or A.D. 476, this is in agreement with the standard *ahargana* used today for the calculations of the *Sūrya-siddhānta*.

For example, October 1, 1965, corresponds to day 1,850,569 in Kali-yuga. On the basis of this information one can calculate that the Kali-yuga began on February 18, 3102 B.C., according to the Gregorian calendar. It is for this reason that Vaiṣṇavas maintain that the pastimes of Kṛṣṇa with the Pāṇḍavas in the battle of Kurukṣetra took place about 5,000 years ago.

Of course, it comes as no surprise that the standard view of Western scholars is that this date for the start of Kali-yuga is fictitious. Indeed, these scholars maintain that the battle of Kurukṣetra itself is fictitious, and that the civilization described in the Vedic literature is simply a product of poetic imagination. It is therefore interesting to ask what modern astronomers have to say about the positions of the planets on February 18, 3102 B.C.

Table 5 lists the longitudes of the planets relative to the reference star Zeta Piscium at the beginning of Kali-yuga. The figures under "Modern True

TABLE 5

The Celestial Longitudes of the Planets at the Start of Kali-yuga

Planet	Modern Mean Longitude	Modern True Longitude
Moon	–6;04	–1;14
Sun	–5;40	–3;39
Mercury	–38;09	–19;07
Venus	27;34	8;54
Mars	–17;25	–6;59
Jupiter	11;06	10;13
Saturn	–25;11	–27;52
Rāhu	–162;44	–162;44

This table shows the celestial longitudes of the planets relative to the star Zeta Piscium (Revatī in Sanskrit) at sunrise on February 18, 3102 B.C., the beginning of Kali-yuga. Each longitude is expressed as degrees; minutes.

Longitude" represent the true positions of the planets at this time according to modern calculations. (These calculations were done with computer programs published by Duffett-Smith (DF).) We can see that, according to modern astronomy, an approximate alignment of the planets did occur at the beginning of Kali-yuga. Five of the planets were within 10° of the Vedic reference star, exceptions being Mercury, at -19°, and Saturn, at -27°. Rāhu was also within 18° of the position opposite Zeta Piscium.

The figures under "Modern Mean Longitude" represent the mean positions of the planets at the beginning of Kali-yuga. The mean position of a planet, according to modern astronomy, is the position the planet would have if it moved uniformly at its average rate of motion. Since the planets speed up and slow down, the true position is sometimes ahead of the mean position and sometimes behind it. Similar concepts of true and mean positions are found in the *Sūrya-siddhānta*, and we note that while the *Sūrya-siddhānta* assumes an exact mean alignment at the start of Kali-yuga, it assumes only an approximate true alignment.

Planetary alignments such as the one in Table 5 are quite rare. To find out how rare they are, we carried out a computer search for alignments by computing the planetary positions at three-day intervals from the start of Kali-yuga to the present. We measured the closeness of an alignment by averaging the absolute values of the planetary longitudes relative to Zeta Piscium. (For Rāhu, of course, we used the absolute value of the longitude relative to a point 180° from Zeta Piscium.) Our program divided the time from the start of Kali-yuga to the present into approximately 510 ten-year intervals. In this entire period we found only three ten-year intervals in which an alignment occurred that was as close as the one occurring at the beginning of Kali-yuga.

We would suggest that the dating of the start of Kali-yuga at 3102 B.C. is based on actual historical accounts, and that the tradition of an unusual alignment of the planets at this time is also a matter of historical fact. The opinion of the modern scholars is that the epoch of Kali-yuga was concocted during the early medieval period. According to this hypothesis, Indian astronomers used borrowed Greek astronomy to determine that a near planetary alignment occurred in 3102 B.C. After performing the laborious calculations needed to discover this, they then invented the fictitious era of Kali-yuga and convinced the entire subcontinent of India that this era had been going on for some three thousand years. Subsequently, many different *Purāṇas* were written in accordance with this chronology, and people all over India became convinced that these works, although unknown to their forefathers, were really thousands of years old.

One might ask why anyone would even think of searching for astronomical alignments over a period of thousands of years into the past and then redefining the history of an entire civilization on the basis of a particular discovered alignment. It seems more plausible to suppose that the story of Kali-yuga is genuine, that the alignment occurring at its start is a matter of historical recollection, and that the *Purāṇas* really were written prior to the beginning of this era.

We should note that many historical records exist in India that make use of dates expressed as years since the beginning of Kali-yuga. In many cases, these dates are substantially less than 3102—that is, they represent times before the beginning of the Christian era. Interesting examples of such dates are given in the book *Ādi Śaṅkara* (AS), edited by S. D. Kulkarni, in connection with the dating of Śaṅkarācārya. One will also find references to such dates in *Age of Bhārata War* (ABW), a series of papers on the date of the *Mahābhārata*, edited by G. C. Agarwala. The existence of many such dates from different parts of India suggests that the Kali era, with its 3102 B.C. starting date, is real and not a concoction of post-Ptolemaic medieval astronomers. (Some references will give 3101 B.C. as the starting date of the Kali-yuga. One reason for this discrepancy is that in some cases a year 0 is counted between A. D. 1 and 1 B.C., and in other cases this is not done.)

At this point the objection might be raised that the alignment determined by modern calculation for the beginning of Kali-yuga is approximate, whereas the astronomical *siddhāntas* generally assume an exact alignment. This seems to indicate a serious defect in the *jyotiṣa śāstras*.

In reply, we should note that although modern calculations are quite accurate for our own historical period, we know of no astronomical observations that can be used to check them prior to a few hundred years B.C. It is therefore possible that modern calculations are not entirely accurate at 3102 B.C. and that the planetary alignment at that date was nearly exact. Of course, if the alignment was as inexact as Table 5 indicates, then it would be necessary to suppose that a significant error was introduced into the *jyotiṣa śāstras*, perhaps in fairly recent times. However, even this hypothesis is not consistent with the theory that 3102 B.C. was selected by Ptolemaic calculations, since these calculations also indicate that a very rough planetary alignment occurred at this date.

Apart from this, we should note that the astronomical *siddhāntas* do not show perfect accuracy over long periods of time. This is indicated by the *Sūrya-siddhānta* itself in the following statement, which a representative of the sun-god speaks to the *asura* Maya:

> O Maya, hear attentively the excellent knowledge of the science of astronomy which the sun himself formerly taught to the great saints in each of the *yugas*.
>
> I teach you the same ancient science. . . . But the difference between the present and the ancient works is caused only by time, on account of the revolution of the *yugas* (SS, p. 2).

According to the *jyotiṣa śāstras* themselves, the astronomical information they contain was based on two sources: (1) revelation from demigods, and (2) human observation. The calculations in the astronomical *siddhāntas* are simple enough to be suitable for hand calculation, but as a result they tend to lose accuracy over time. The above statement by the sun's representative indicates that these works were updated from time to time in order to keep them in agreement with celestial phenomena.

We have made a computer study comparing the *Sūrya-siddhānta* with modern astronomical calculations. This study suggests that the *Sūrya-siddhānta* was probably updated some time around A.D. 1000, since its calculations agree most closely with modern calculations at that time. However, this does not mean that this is the date when the *Sūrya-siddhānta* was first written. Rather, the parameters of planetary motion in the existing text may have been brought up to date at that time. Since the original text was as useful as ever once its parameters were updated, there was no need to change it, and thus it may date back to a very remote period.

A detailed discussion concerning the date and origin of Āryabhaṭa's astronomical system is found in Appendix 2. There we observe that the parameters for this astronomical system were probably determined by observation during Āryabhaṭa's lifetime, in the late 5th and early 6th centuries A.D. Regarding his theoretical methods, Āryabhaṭa wrote, "By the grace of Brahmā the precious sunken jewel of true knowledge has been brought up by me from the ocean of true and false knowledge by means of the boat of my own intellect" (VW, p. 213). This suggests that Āryabhaṭa did not claim to have created anything new. Rather, he simply reclaimed old knowledge that had become confused in the course of time.

In general, we would suggest that revelation of astronomical information by demigods was common in ancient times prior to the beginning of Kali-yuga. In the period of Kali-yuga, human observation has been largely used to keep astronomical systems up to date, and as a result, many parameters in existing works will tend to have a fairly recent origin. Since the Indian astronomical tradition was clearly very conservative and was mainly oriented towards fulfilling customary day-to-day needs, it is quite possible that the methods used in these works are extremely ancient.

As a final point, we should consider the objection that Indian astronomers have not given detailed accounts of how they made observations or how they computed their astronomical parameters on the basis of these observations. This suggests to some that a tradition of sophisticated astronomical observation never existed in India.

One answer to this objection is that there is abundant evidence for the existence of elaborate programs of astronomical observation in India in recent centuries. The cover of this book depicts an astronomical instrument seen in Benares in 1772 by an Englishman named Robert Barker; it was said to be about 200 years old at that time. About 20 feet high, this structure includes two quadrants, divided into degrees, which were used to measure the position of the sun. It was part of an observatory including several other large stone and brass instruments designed for sighting the stars and planets (PR, pp. 31–33).

Similar instruments were built in Agra and Delhi. The observatory at Delhi was built by Rajah Jayasingh in 1710 under the auspices of Mohammed Shah, and it can still be seen today. Although these observatories are quite recent, there is no reason to suppose that they first began to be built a few centuries ago. It is certainly possible that over a period of thousands of years such observatories were erected in India when needed.

The reason we do not find elaborate accounts of observational methods in the *jyotiṣa śāstras* is that these works were intended simply as brief guides for calculators, not as comprehensive textbooks. Textbooks were never written, since it was believed that knowledge should be disclosed only to qualified disciples. This is shown by the following statement in the *Sūrya-siddhānta*:

"O Maya, this science, secret even to the Gods, is not to be given to anybody but the well-examined pupil who has attended one whole year" (SS, p. 56). Similarly, after mention of a motor based on mercury that powers a revolving model of the universe, we find this statement: "The method of constructing the revolving instrument is to be kept a secret, as by diffusion here it will be known to all" (SS, p. 90). The story of the false disciple of Droṇācārya in the *Mahābhārata* shows that this restrictive approach to the dissemination of knowledge was standard in Vedic culture.

1.E. THE DISTANCES AND SIZES OF THE PLANETS

In Section 1.a we derived relative distances between the planets from the orbital data contained in the *Sūrya-siddhānta*. These distances are expressed in units of the earth-sun distance, or AU. In this section we will consider absolute distances measured in miles or *yojanas* and point out an interesting feature of the *Sūrya-siddhānta*: it seems that figures for the diameters of the planets are encoded in a verse in the seventh chapter of this text. These diameters agree quite well with the planetary diameters determined by modern astronomy. This is remarkable, since it is hard to see how one could arrive at these diameters by observation without the aid of powerful modern telescopes.

Absolute distances are given in the *Sūrya-siddhānta* in *yojanas*—the same distance unit used throughout the *Śrīmad-Bhāgavatam*. To convert such a unit into Western units such as miles or kilometers, it is necessary to find some distances that we can measure today and that have also been measured in *yojanas*. Śrīla Prabhupāda has used a figure of eight miles per *yojana* throughout his books, and this information is presumably based on the joint usage of miles and *yojanas* in India.

Since some doubt has occasionally been expressed concerning the size of the *yojana*, here is some additional information concerning the definition of this unit of length. One standard definition of a *yojana* is as follows: one *yojana* equals four *krośas*, where a *krośa* is the maximum distance over which a healthy man can shout and be heard by someone with good hearing (AA). It is difficult to pin down this latter figure precisely, but it surely could not be much over two miles. Another definition is that a *yojana* equals 8,000 *nr*, or heights of a man. Using 8 miles per *yojana* and 5,280 feet per mile, we obtain 5.28 feet for the height of a man, which is not unreasonable. In Appendix 1 we give some other definitions of the *yojana* based on the human body.

A more precise definition of a *yojana* can be obtained by making use of the figures for the diameter of the earth given by Indian astronomers. Āryabhaṭa gives a figure of 1,050 *yojanas* for the diameter of the earth (AA). Using the current figure of 7,928 miles for the diameter of the earth, we obtain 7,928/1,050

= 7.55 miles per *yojana*, which is close to 8. We also note that Alberuni (AL, p. 167) gives a figure of 8 miles per *yojana*, although it is not completely clear whether his mile is the same as ours.

In the *Siddhānta-śiromaṇi* of Bhāskarācārya, the diameter of the earth is given as 1,581 *yojanas* (SSB2, p. 83), and in the *Sūrya-siddhānta* a diameter of 1,600 *yojanas* is used (SS, p. 11). These numbers yield about 5 miles per *yojana*, which is too small to be consistent with either the 8 miles per *yojana*

TABLE 6

The Diameters of the Planets, According to the Sūrya-siddhānta

Planet	Orbit	Reduced Diameter	SS Diameter Yojanas	SS Diameter Miles	W Diameter Miles	W/SS
Moon	324,000	480.00	480.00	2,400.0	2,160	.90
Sun	4,331,500	486.21	6,500.00	32,500.0	865,110	26.62
Mercury	4,331,500	45.00	601.60	3,008.0	3,100	1.03
Venus	4,331,500	60.00	802.13	4,010.6	7,560	1.89
Earth	0	—	1,600.00	8,000.0	7,928	.99
Mars	8,146,909	30.00	754.34	3,771.7	4,191	1.11
Jupiter	51,375,764	52.50	8,324.80	41,624.0	86,850	2.09
Saturn	127,668,255	37.50	14,776.00	73,882.0	72,000	.97

The first column lists the planetary orbital circumferences in *yojanas* (SS, p. 87). The second column lists the diameters of the planets in *yojanas* reduced to the orbit of the moon (SS, p. 59). The third column lists the corresponding actual diameters (in *yojanas* and miles). Except for the sun, moon, and earth (where the figures are taken from SS, p. 41), these values are computed using the data in columns 1 and 2. The fourth column lists the current Western values for the planetary diameters, and the last column lists the ratios between the Western diameters and the diameters based on the *Sūrya-siddhānta*.

or the 8,000 *nṛ* per *yojana* standards. (At 5 miles per *yojana* we obtain 3.3 feet for the height of a man, which is clearly too short.) The Indian astronomer Parameśvara suggests that these works use another standard for the length of a *yojana*, and this is borne out by the fact that their distance figures are consistently 60% larger than those given by Āryabhaṭa. Thus, it seems clear that a *yojana* has traditionally represented a distance of a few miles, with 5 and approximately 8 being two standard values used by astronomers.

At this point it is worthwhile considering how early Indian astronomers obtained values for the diameter of the earth. The method described in their writings (GP, p. 84) is similar to the one reportedly used by the ancient Greek astronomer Eratosthenes. If the earth is a sphere, then the vertical directions at two different points should differ in angle by an amount equal to 360 times the distance between the points divided by the circumference of the earth. This angle can be determined by measuring the tilt of the noon sunlight from vertical at one place, and simultaneously measuring the same tilt at the other place (assuming that the sun's rays at the two places run parallel to one another). At a separation of, say, 500 miles, the difference in tilt angles should be about 7°, a value that can be easily measured and used to compute the earth's circumference and diameter.

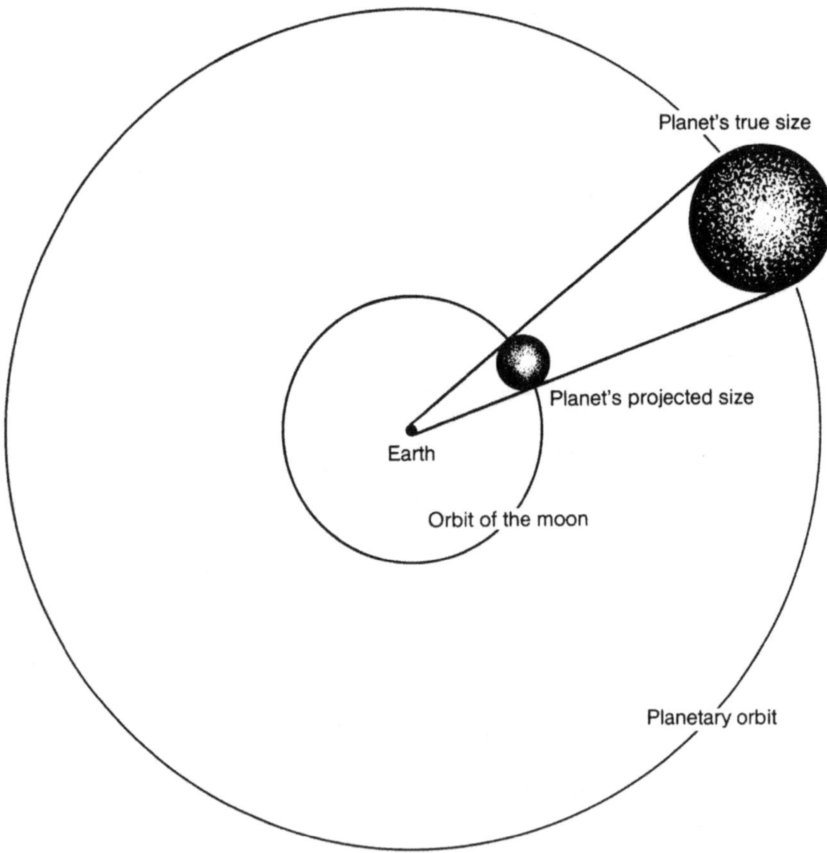

Figure 2. The relationship between the true diameter of a planet and its diameter projected to the orbit of the moon.

The *Sūrya-siddhānta* lists the diameter of the moon as 480 *yojanas* and the circumference of the moon's orbit as 324,000 *yojanas*. If we convert these figures into miles by multiplying by the *Sūrya-siddhānta* value of 5 miles per *yojana*, we obtain 2,400 and 1,620,000. According to modern Western figures, the diameter of the moon is 2,160 miles, and the circumference of the moon's orbit is 2π times the earth-to-moon distance of 238,000 miles, or 1,495,000 miles. Thus the *Sūrya-siddhānta* agrees closely with modern astronomy as to the size of the moon and its distance from the earth.

Table 6 lists some figures taken from the *Sūrya-siddhānta* giving the circumferences of the orbits of the planets (with the earth as center), and the diameters of the discs of the planets themselves. The orbital circumferences of the planets other than the moon are much smaller than they should be according to modern astronomy.

The diameter of the moon is also the only planetary diameter that seems, at first glance, to agree with modern data. Thus, the diameter given for the sun is 6,500 *yojanas*, or 32,500 miles, whereas the modern figure for the diameter of the sun is 865,110 miles. The diameter figures for Mercury, Venus, Mars, Jupiter, and Saturn are given in *yojanas* for the size of the planetary disc when projected to the orbit of the moon (see Figure 2). These figures enable us to visualize how large the planets should appear in comparison with the full moon. On the average the figures are too large by a factor of ten, and they imply that we should easily be able to see the discs of the planets with the naked eye. Of course, without the aid of a telescope, we normally see these planets as starlike points.

The discs of the planets Mercury through Saturn actually range from a few seconds of arc to about 1', and for comparison the disc of the full moon covers about 31.2' of arc. This means that a planetary diameter projected to the orbit of the moon should be no greater than 15.4 *yojanas*. From the standpoint of modern thought, it is not surprising that an ancient astronomical work like the *Sūrya-siddhānta* should give inaccurate figures for the sizes of the planetary discs. In fact, it seems remarkable that ancient astronomers lacking telescopes could have seen that the planets other than the sun and moon actually have discs.

If we look more closely at the data in Table 6, however, we can make a very striking discovery. Since the diameters of Mercury through Saturn are projected on the orbit of the moon, their real diameters should be given by the formula:

$$\text{real diameter} = \frac{\text{projected diameter} \times \text{orbital circumference}}{\text{moon's orbital circumference}}$$

If we compute the real diameters using this formula and the data in Table 6, we find that the answers agree very well with the modern figures for the diameters of the planets (see the last three columns of the table). Thus, the distance figures and the values for the projected (or apparent) diameters disagree with modern

TABLE 7

Modern Values for Planetary Distances and Diameters vs. Those of the Sūrya-siddhānta

Planet	Mean Distance from Earth	Apparent Diameter	Real Diameter
Moon	agrees	agrees	agrees
Sun	disagrees	agrees	disagrees
Mercury	disagrees	disagrees	agrees
Venus	disagrees	disagrees	off by ½
Earth	—	—	agrees
Mars	disagrees	disagrees	agrees
Jupiter	disagrees	disagrees	off by ½
Saturn	disagrees	disagrees	agrees

The entry "agrees" means that the *Sūrya-siddhānta* value falls within about 10% of the modern value. The cases that are "off by ½" fall within less than 7% of the modern values after being doubled.

astronomy, but the actual diameters implied by these figures agree. This is very surprising indeed, considering that modern astronomers have traditionally computed the planetary diameters by using measured values of distances and apparent diameters.

We note that the diameters computed for Mercury, Mars, and Saturn using our formula are very close to the modern values, while the figures for Venus and Jupiter are off by almost exactly ½. This is an error, but we suggest that it is not simply due to ignorance of the actual diameters of these two planets. Rather, the erroneous factor of ½ may have been introduced when a careless copyist mistook "radius" for "diameter" when copying an old text that was later used in compiling the present *Sūrya-siddhānta*.

This explanation is based on the otherwise excellent agreement that exists between the *Sūrya-siddhānta* diameters and modern values, and on our hypothesis that existing *jyotiṣa śāstras* such as the *Sūrya-siddhānta* may be imperfectly preserved remnants of an older Vedic astronomical science. We suggest that accurate knowledge of planetary diameters existed in Vedic times, but that this knowledge was garbled at some point after the advent of Kali-yuga. However, this knowledge is still present in an encoded form in the present text of the *Sūrya-siddhānta*.

The circumferences of the planetary orbits listed in Table 6 are based on the theory of the *Sūrya-siddhānta* that all planets move through space with the same *average* speed. Using this theory, one can compute the average distances of the planets from their average apparent speeds, and this is how the circumferences listed in Table 6 were computed in the *Sūrya-siddhānta*. The same theory concerning the motions of the planets can be found in other works of the siddhāntic school, but it is not mentioned in the *Śrīmad-Bhāgavatam*. This theory disagrees with that of modern astronomers, who maintain that the planets move more slowly the further they are from the sun.

We should emphasize that this theory applies only to the planets' average speeds in circular motion around the earth. The actual speeds of the planets vary in the *Sūrya-siddhānta*, and a rule is given for computing the change in apparent diameter of the planets as their distance from the earth changes. The motions of the planets are said to be caused by the *pravaha* wind and by the action of reins of wind pulled by demigods.

Since the relative distances of the planets derived from the *Sūrya-siddhānta* in Section 1.a are not consistent with the orbital circumferences listed in Table 6, it would seem that the *Sūrya-siddhānta* contains material representing more than one theoretical viewpoint. This also makes sense if we suppose that the surviving *jyotiṣa śāstras* may represent the incompletely understood remnants of a body of knowledge that was more complete in the ancient past.

Table 7 sums up our observations on the diameters and distances of the planets given in the *Sūrya-siddhānta*. At present we have no explanation of how diameters agreeing so closely with modern values were found, even though estimates of distances and apparent diameters disagree. According to current astronomical thinking, the real diameters can be obtained only by making measurements using powerful telescopes and then combining these results with accurate knowledge of the planetary distances. However, other methods may have been available in Vedic times.

We should note, by the way, that the numbers for planetary diameters can be found not only in our English translation of the *Sūrya-siddhānta* (SS), but also in Śrīla Bhaktisiddhānta Sarasvatī Ṭhākura's Bengali translation. This strongly indicates that these numbers belong to the original *Sūrya-siddhānta*, and were not inserted as a hoax in recent times.

We should also consider the possibility that the planetary diameters given in the *Sūrya-siddhānta* were derived from Greek sources. It turns out that there is a medieval tradition regarding the distances and diameters of the planets that can be traced back to a book by Ptolemy entitled *Planetary Hypotheses*. In this book the apparent diameters of the planets are given as fractions of the sun's apparent diameter. For the moon, Mercury, Venus, Mars, Jupiter, and Saturn, these apparent diameters are stated by Ptolemy to be, respectively, $1\frac{1}{3}$, $\frac{1}{15}$, $\frac{1}{10}$,

1/20, 1/12, and 1/18 (SW, p. 167). Corresponding apparent diameters can be computed from the *Sūrya-siddhānta* data by taking the diameters of the planets reduced to the moon's orbit and dividing by 486.21, the diameter of the sun reduced to the moon's orbit. The values obtained, however, are quite different from Ptolemy's apparent diameters.

Ptolemy also computes actual diameters, expressed as multiples of the earth's diameter, using his apparent diameters and his values for the average distances of the planets from the earth. We have converted his actual diameters into miles by multiplying them by 7,928 miles, our modern value for the diameter of the earth. The results for the moon, Mercury, Venus, Mars, Jupiter, and Saturn are 2,312, 294, 2,246, 9,061, 34,553, and 34,090, respectively. (See SW, p. 170.) Apart from the figure for the moon, these diameters show no relationship with either the modern planetary diameters or the diameters obtained from the *Sūrya-siddhānta* and listed in Table 6.

The only feature that the *Sūrya-siddhānta* and Ptolemy seem to share with regard to the diameters of the planets is that both give unrealistically large values for apparent diameters. If the planets actually had such large apparent diameters, they would appear to the naked eye as clearly visible discs rather than as stars. The ancient planetary diameters would therefore seem to be completely fictitious, were it not for the fact that in the case of the *Sūrya-siddhānta*, they correspond to realistic, actual diameters as seen from unrealistically short distances.

1.F. THE SIZE OF THE UNIVERSE

In the *Śrīmad-Bhāgavatam* a figure of 500 million *yojanas* is given for the diameter of the universe. On the basis of 8 miles per *yojana*, this comes to 4 billion miles, a distance that can accommodate the orbit of Saturn (according to modern distance figures), but that is smaller than the orbital diameters of Uranus, Neptune, and Pluto. Since this figure for the diameter of the universe seems to be quite small, it is interesting to note the purport given by Śrīla Prabhupāda to CC ML 21.84:

> [Text:] Kṛṣṇa said, "Your particular universe extends four billion miles; therefore it is the smallest of all the universes. Consequently you have only four heads."
>
> [Purport:] Śrīla Bhaktisiddhānta Sarasvatī Ṭhākura, one of the greatest astrologers of his time, gives information from *Siddhānta-śiromaṇi* that this universe measures 18,712,069,200,000,000 × 8 miles. This is the circumference of this universe. According to some, this is only half the circumference.

In his *Anubhāṣya* commentary on this verse of *Caitanya-caritāmṛta*, Śrīla Bhaktisiddhānta Sarasvatī quotes from *Sūrya-siddhānta* 12.90, "The circumference of the sphere of the Brahmāndee in which the sun's rays spread is 18,712,080,864,000,000 *yojanas*" (SS, p. 87). Then he quotes *Siddhānta-śiromaṇi*, *Golādhyāya Bhuvana-kośa*: "Some astronomers have asserted the circumference of the circle of heaven to be 18,712,069,200,000,000 *yojanas* in length. Some say that this is the length of the zone binding the two hemispheres of the Brahmāṇḍa. Some Paurāṇikas say that this is the length of the circumference of the Lokāloka Parvata [*adṛśya-dṛśyaka-girim*]" (SSB1, p. 126).

Here the circumference of 18,712,069,200,000,000 *yojanas* corresponds to a diameter of 5,956,200,000,000,000 *yojanas*. This number is much larger than the 500,000,000-*yojana* diameter given in the *Bhāgavatam*, and we might ask how it relates to it. According to the *Bhāgavatam* (5.20.37),

> By the supreme will of Kṛṣṇa, the mountain known as Lokāloka has been installed as the outer border of the three worlds—Bhūrloka, Bhuvarloka and Svarloka—to control the rays of the sun throughout the universe. All the luminaries, from the sun up to Dhruvaloka, distribute their rays throughout the three worlds, but only within the boundary formed by this mountain.

This verse reconciles the statement that the 18-quadrillion-*yojana* circumference is the limit of distribution of the sun's rays with the statement that it is the circumference of Lokāloka Mountain. We also note that in SB 5.20.38 the diameter of Lokāloka Mountain is stated to be half the diameter of the universe. This is consistent with the statement in Śrīla Prabhupāda's purport that "according to some, this is only half the circumference." We are thus left with a picture of the universe in which the rays of the sun and other luminaries spread to a radial distance of 2,978,100,000,000,000 *yojanas*, and are there blocked in all directions by an enormous mountain. This mountain lies halfway between the sun and the beginning of the outer coverings of the universe. This means that the distance from the sun to the coverings of the universe is some 5,077 light-years, where a light-year is the distance traveled in one year by a beam of light moving at 186,000 miles per second and we use the *Sūrya-siddhānta*'s 5-mile *yojanas*.

In Chapters 3 and 4 we will say more about the possible relation between this very large universal radius and the much smaller figure given in the *Bhāgavatam*. At present we will consider what the *jyotiṣa śāstras* have to say about the radius of the universe. It turns out that the *Siddhānta-śiromaṇi*, the *Sūrya-siddhānta*, and many other *jyotiṣa śāstras* give a simple rule for computing this number.

The *Sūrya-siddhānta* gives the following rule: "Multiply the number of . . . revolutions of the moon in a *kalpa* by the moon's orbit . . . : the product is equal to the orbit of heaven (or the circumference of the middle of the *brahmāṇḍa*): to this orbit the sun's rays reach" (SS, p. 86). If we perform this calculation, we find that the circumference of the *brahmāṇḍa*, or universe, is:

57,753,336 × 1,000 × 324,000 = 18,712,080,864,000,000 *yojanas*

In *The Aryabhatiya of Aryabhata* we find the statement that the circumference of the sky (*ākāśa-kakṣa*) in *yojanas* is equal to 10 times the number of minutes of arc covered by the moon during one *divya-yuga* (AA, p. 13). This comes to:

57,753,336 × 360 × 60 × 10 = 12,474,720,576,000 *yojanas*

When interpreting this figure, we should keep in mind that Āryabhaṭa used a *yojana* of about 7.55 miles rather than 5 miles. If we convert Āryabhaṭa's figure to 5-mile *yojanas*, we obtain a universal circumference that is almost exactly one thousandth of the figure cited in *Sūrya-siddhānta* and *Siddhānta-śiromaṇi*. The reason for this is that Āryabhaṭa used the number of revolutions of the moon in a *divya-yuga* rather than the number of revolutions in a *kalpa*. (There are 1,000 *divya-yugas* per *kalpa*.)

We mention Āryabhaṭa's calculation for the sake of completeness. There are a number of ways in which Āryabhaṭa differs from other Indian astronomers (AA). For example, he is unique in making the four *yugas* equal in length, and he also suggests that the earth rotates daily on its axis. (All other Indian astronomers speak of the *kāla-cakra* rotating around a fixed earth.) Our main point here is that very large figures for the size of the universe were commonly presented in the *jyotiṣa śāstras*, and such figures have been accepted by Śrīla Bhaktisiddhānta Sarasvatī Ṭhākura and Śrīla Prabhupāda.

Vedic Physics: The Nature of Space, Time, and Matter

"By Him even the great sages and demigods are placed into illusion, as one is bewildered by the illusory representations of water seen in fire, or land seen on water. Only because of Him do the material universes, temporarily manifested by the reactions of the three modes of nature, appear factual, although they are unreal" (SB 1.1.1).

Our ideas of the nature of space, time, and matter are essential ingredients in our understanding of the cosmos. When we look into the heavens, our direct sensory data consist of patterns of light. These patterns say nothing, in and of themselves, about the nature of the sources of this light. In order to say something about the cosmic manifestations that have produced the light, it is necessary to assume that the universe is made of some kind of stuff, or matter, that has certain characteristics and obeys certain laws. Given such assumptions, we can then ask ourselves what arrangement of this matter, acting in accordance with the laws, would produce the observed light patterns. If we are successful in putting together a consistent explanation of the observed data based on the assumed laws and properties, then we tend to suppose that we have correctly understood the structure of the universe. In our mind's eye, our theoretical models take on an air of concrete reality, and it almost seems as though we were holding the universe in the palm of our hand.

Throughout most of modern human history, people have been limited to the surface of the earth, and they have based their ideas of the nature of matter on observations that we can perform in this limited domain using our ordinary senses. Over the last two or three hundred years, Western scientists have used experimental observation and the analysis of experimental results to build up an extensive body of knowledge—the science of modern physics—which gives a detailed picture of the properties of matter and the laws governing its behavior. The modern Western understanding of the nature and structure of the universe as a whole is based on interpreting observed celestial phenomena within the framework of modern physics.

The thesis of this book is that the framework of modern physics is too limited to accommodate many phenomena that occur within this universe. In particular, this framework cannot accommodate many features of the universe that are described in the Vedic literature, and thus the Vedic accounts often seem absurd or mythological when viewed from the perspective of modern science. At the present time, certain assumptions of modern physics have been adopted by people in general as the very foundation of their world view. These assumptions are incompatible with the underlying assumptions of the Vedic world view, and thus they tend to block people from having free access to the Vedic literature. In this section we will try to alleviate this difficulty by discussing the nature of the material energy as described in the Vedic literature. Since this is a very deep and complex subject, we will be able to touch on only a few points that are relevant to the understanding of Vedic cosmology.

2.A. EXTENDING OUR PHYSICAL WORLD VIEW

Before making a truly radical departure from our familiar conceptions, we will begin by discussing some relatively moderate instances in which the Vedic literature refers to phenomena and theoretical ideas that do not fit into the current framework of scientific thought. These examples illustrate two main points: (1) Although many Vedic ideas contradict current scientific thinking, they also allow for the possibility that the contradictions can be alleviated by extending the conceptual scope of modern science. (2) Many ideas relevant to our physical world-picture are alluded to only briefly in works such as the *Śrīmad-Bhāgavatam*, since these works were not intended to serve as textbooks of astronomy or physical science. Thus the conceptual advances needed to reconcile the Vedic world view with modern science may be difficult to make, since they require ideas that radically extend current theories but are not explicitly spelled out in available Vedic texts.

Our first example is found in SB 3.26.34p. There we read that the ethereal element provides a substrate for the production of subtle forms by the mind, and that it is also involved in the circulation of vital air within the body. Śrīla Prabhupāda indicates that "this verse is the potential basis of great scientific research work," and indeed, it provides a clear idea of how the subtle mind may interact with the gross elements of the body and brain.

In the theoretical structure of modern physics, however, there is at present no place for such a conception of the mind and the ethereal element (although some physicists have tentatively begun to entertain such ideas). As a consequence, scientists still generally adhere to the idea that it is impossible for the brain to interact with a distinct nonphysical mind. This in turn makes it impossible for them to give credence to many phenomena that imply the existence of such a mind, even though empirical evidence for these phenomena has existed for many years. These phenomena include the psychic events studied by the parapsychologists, out-of-body experiences, and the spontaneous remembrance of previous incarnations by small children.

It is not our purpose here to make a case for the reality of such phenomena. Our main point is that it is very difficult for people (including scientists) to seriously contemplate particular ideas about reality unless those ideas fit neatly into a familiar and accepted conceptual system. The current theories of physics have been worked out in great technical detail, and one who lives in the conceptual universe these theories provide may find that the Vedic idea of ether seems crude and unimpressive. Openness to the Vedic ideas may also be blocked by certain misconceptions, such as the idea that ether must be like the "luminiferous ether" rejected by Einstein. Yet the possibility nonetheless exists that physical theory can be extended by introducing a new conception of the ether that agrees with the Vedic conception and is consistent with experimental observations. And such an extended theory may provide explanations for many phenomena presently considered scientifically impossible.

Texts such as the *Śrīmad-Bhāgavatam* were written for the purpose of clearly explaining certain spiritual ideas to people in general. However, they inevitably make reference to many other ideas that were familiar to people of the ancient Vedic culture but that may be very unfamiliar to people of modern Western background. One interesting example is the analogy given by Śrīla Sanātana Gosvāmī in which the transformation of a lowborn man into a *brāhmaṇa* is compared to the transformation of bell metal into gold by an alchemical process (SB 5.24.17p).

The alchemical process itself is not described, and on the basis of modern science we might tend to regard such a transformation as impossible. Yet the dictionary defines bell metal as an alloy of copper and tin, and if we consult the periodic table of the elements, we find that the atomic numbers of copper and tin added together give the atomic number of gold. This suggests that there just might be something to this example, but if so, it clearly involves an extensive body of practical and theoretical knowledge that is completely unknown to us. For Sanātana Gosvāmī, however, this transformation simply provided a familiar example to illustrate a point about the spiritual transformation of human beings.

2.B. THE POSITION OF KṚṢṆA

Thus far, we have discussed Vedic references to phenomena and theoretical entities that do not fit into the rigorously defined theories of modern physics but that can be readily inserted into our ordinary picture of the world around us. In this book, however, we will be dealing with many things that do not seem to be at all compatible with that picture. We suggest that to accommodate these things, it is necessary for us to re-examine our basic ideas concerning the nature of space.

Modern physics and astronomy began with the idea that matter is made of tiny bits of substance, each of which has a location in three-dimensional space. According to this idea, which was strongly developed by Descartes and Newton, three-dimensional space can be seen as an absolute, pre-existing container in which all material events take place. This idea is quite consistent with the picture of the world provided by our own senses, and it tends to provide an unquestioned background for all of our thinking. However, many cultures have maintained quite different ideas about the nature of space, and this is also true of the Vedic culture.

To understand the Vedic conception of space, it is necessary to consider the position of Kṛṣṇa as the absolute cause of all causes. Clearly we cannot regard the transcendental form of Kṛṣṇa as being composed of tiny bits of substance situated at different locations in three-dimensional space. Whether we regard the tiny bits as spiritual or material, such a form would certainly be limited and relative. The actual nature of Kṛṣṇa's form is indicated by the following verses from the *Brahma-saṁhitā:*

> I worship Govinda, the primeval Lord, whose transcendental form is full of bliss, truth, and substantiality and is thus full of the most dazzling splendor. Each of the limbs of that transcendental figure possesses in itself the full-fledged functions of all the organs, and He eternally sees, maintains, and manifests the infinite universes, both spiritual and mundane [SBS 5.32].
>
> He is an undifferentiated entity, as there is no distinction between the potency and the possessor thereof. In His work of creation of millions of worlds, His potency remains inseparable. All the universes exist in Him, and He is present in His fullness in every one of the atoms that are scattered throughout the universe, at one and the same time. Such is the primeval Lord whom I adore [SBS 5.35].

These verses indicate that the form of Kṛṣṇa is made of many parts, but that each part is identical to the whole. Also, all space is within the form of Kṛṣṇa, but at the same time Kṛṣṇa is fully present within every atom. One implication of this is that the entire universe, which is within Kṛṣṇa, is fully present within every atom of the universe. Such a state of affairs cannot be visualized in three-dimensional terms, and indeed, it is not possible within three-dimensional space. The statement that reality is like this must simply be taken as an axiom describing the position of Kṛṣṇa as the Supreme Absolute Truth. Thus, the Vedic concept of space begins with a statement of Kṛṣṇa's unified nature, rather than with the geometric axioms defining three-dimensional space.

Here we will introduce an idea of higher-dimensional space that may help us understand the ideas about space implicit in the Vedic literature. The term *higher-dimensional* is borrowed from modern mathematics; it does not appear directly in Vedic literature. It is part of an attempt to bridge the conceptual gap between modern thinking and the Vedic world view. Naturally, since the traditional followers of Vedic culture have not been confronted with such a gap, they have not been motivated to introduce ideas to bridge it.

The most fundamental feature of the Vedic idea of space is that many more things can be brought close together in this space than the geometric rules of three-dimensional space allow. In the course of this chapter we will give several examples from the Vedic literature illustrating this theme. Since the higher-dimensional spaces of mathematics also permit more things to be brought together than the rules of three-dimensional space allow, we have chosen the term *higher-dimensional* to refer to this feature of the Vedic view of reality.

Although Kṛṣṇa's situation is very difficult for us to visualize, we can nonetheless understand from Vedic statements describing Kṛṣṇa that space must be higher-dimensional. Kṛṣṇa's situation is that He has full access to every location simultaneously. In ordinary, three-dimensional space we have access, through the operation of our senses of action and perception, to locations within a limited neighborhood, and we can change that neighborhood by moving from one place to another. Thus our situation can be viewed as a restricted form of Kṛṣṇa's situation. A higher-dimensional space corresponds to a situation in which access between locations is more restricted than it is for Kṛṣṇa but less restricted than it is for beings experiencing three-dimensional space.

This concept of higher-dimensional space is closely tied together with the idea of varying levels of sensory development in sentient beings. Access between locations depends on the operation of senses of action and senses of perception, and thus it should be possible in principle to enlarge the space of one's experience by increasing the scope of one's sensory powers.

These ideas about space and its relation to sense perception are implicit in the Vedic literature, and they can best be understood by giving some specific examples. The nature of Kṛṣṇa's absolute position is nicely illustrated by the following story of a visit by Lord Brahmā to Kṛṣṇa in Dvārakā. In the story, Kṛṣṇa first responds to Brahmā's request to see Him by having His secretary ask, "Which Brahmā wishes to see Me?" Brahmā later begins his conversation with Kṛṣṇa by asking why Kṛṣṇa made this inquiry:

> "Why did you inquire which Brahmā had come see You? What is the purpose of such an inquiry? Is there any other Brahmā besides me within this universe?"

> Upon hearing this, Śrī Kṛṣṇa smiled and immediately meditated. Unlimited Brahmās arrived instantly. These Brahmās had different numbers of heads. Some had ten heads, some twenty, some a hundred, some a thousand, some ten thousand, some a hundred thousand, some ten million, and others a hundred million. No one can count the number of faces they had.
>
> There also arrived many Lord Śivas with various heads numbering one hundred thousand and ten million. Many Indras also arrived, and they had hundreds of thousands of eyes all over their bodies.
>
> When the four-headed Brahmā of this universe saw all these opulences of Kṛṣṇa, he became very bewildered and considered himself a rabbit among many elephants.
>
> All the Brahmās who came to see Kṛṣṇa offered their respects at His lotus feet, and when they did this, their helmets touched His lotus feet. No one can estimate the inconceivable potency of Kṛṣṇa. All the Brahmās who were there were resting in the one body of Kṛṣṇa. When all the helmets struck together at Kṛṣṇa's lotus feet, there was a tumultuous sound. It appeared that the helmets themselves were offering prayers unto Kṛṣṇa's lotus feet.
>
> With folded hands, all the Brahmās and Śivas began to offer prayers unto Lord Kṛṣṇa, saying, "O Lord, You have shown me a great favor. I have been able to see Your lotus feet."
>
> Each of them then said, "It is my great fortune, Lord, that You have called me, thinking of me as Your servant. Now let me know what Your order is so that I may carry it on my heads."
>
> Lord Kṛṣṇa replied, "Since I wanted to see all of you together, I have called all of you here. All of you should be happy. Is there any fear of the demons?"
>
> They replied, "By Your mercy, we are victorious everywhere. Whatever burden there was upon the earth You have taken away by descending on that planet."
>
> This is the proof of Dvārakā's opulence: all the Brahmās thought, "Kṛṣṇa is now staying in my jurisdiction." Thus the opulence of Dvārakā was perceived by each and every one of them. Although they were all assembled together, no one could see anyone but himself.
>
> Lord Kṛṣṇa then bade farewell to all the Brahmās there, and after offering their obeisances, they all returned to their respective homes [CC ML 21.65–80].

In this story it is significant that each of the Brahmās remained within his own universe. This means that Kṛṣṇa was simultaneously manifesting His Dvārakā pastimes in all of those universes. Each Brahmā except ours thought that he was alone with Kṛṣṇa in Dvārakā within his own universe, but by Kṛṣṇa's grace our Brahmā could simultaneously see all the others. This illustrates that Kṛṣṇa has access to all locations at once, and it also shows that, by Kṛṣṇa's grace, different living beings can be given different degrees of spatial access, either permanently or temporarily.

Arjuna's vision of Kṛṣṇa's universal form on the battlefield of Kurukṣetra is another example of Kṛṣṇa's expanding the sensory powers of a living being and giving him access to regions of the universe previously unknown to him. Before revealing this form to Arjuna, Kṛṣṇa said,

> O best of the Bhāratas, see here the different manifestations of Ādityas, Vasus, Rudras, Aśvinī-kumāras, and all the other demigods. Behold the many wonderful things that no one has ever seen or heard of before.
>
> O Arjuna, whatever you want to see, behold at once in this body of Mine! This universal form can show you whatever you now desire to see and whatever you may want to see in the future. Everything—moving and nonmoving—is here completely, in one place [BG 11.6–7].

Thus from one place Arjuna was able to see many different realms occupied by demigods and other kinds of living beings. To perceive such a vast variety of scenes simultaneously, Arjuna clearly had to transcend the limitations of three-dimensional space, and it is significant that Kṛṣṇa made this possible through the medium of His all-pervading universal form. The story of mother Yaśodā's seeing the entire universe (including herself and Kṛṣṇa) within Kṛṣṇa's mouth is another example showing that Kṛṣṇa can reveal all locations through His all-encompassing form (see KB, pp. 83–84).

It is interesting to note that the Brahmās visiting Kṛṣṇa had varying numbers of heads, ranging from four to hundreds of millions. It is rather difficult to understand how millions of heads could be arranged on one body in three-dimensional space, and it is also difficult to see how millions of Brahmās could all be seen simultaneously within one room. We suggest that these things are made possible by the fact that the underlying space is not three-dimensional.

Similar observations could be made about the incident in which Bāṇāsura used 1,000 arms to work 500 bows and shoot 2,000 arrows at a time at Kṛṣṇa. In this case we are dealing with a materially embodied being living on the earth. One might wonder how 500 material arms could be mounted on one shoulder without interfering with one another. And if this is possible, how could they aim 500 bows in the same direction at once? (Did the bows pass through each other?) We suggest that stories of this kind implicitly require higher-dimensional conceptions of space.

We can sum up the idea of dimensionality of space by saying that the greater the degree of access between locations, the higher the dimensionality of the space. Since Kṛṣṇa has simultaneous access to all locations, He perceives space at the highest level of dimensionality. Different living beings will perceive space at different levels of dimensionality, and thus they will have access to different sets of locations (or *lokas*).

It is interesting to note that the idea of higher-dimensional access between locations is a key feature of quantum mechanics. The quantum mechanical atom cannot be represented in three-dimensional space. In fact, to represent something as commonplace as an atom of carbon, quantum mechanics makes use of a kind of infinite-dimensional space called Hilbert space. The three-dimensional bonding of carbon and other atoms is made possible by the higher-dimensional interactions within the atoms. Thus, although the idea of higher-dimensional realms may seem to be an extreme departure from accepted scientific thinking, it is possible to interpret modern physics as laying the groundwork for such an idea.

2.C. MYSTIC SIDDHIS

The eight mystic *siddhis* directly illustrate that sentient beings can operate at different levels of sensory power by being endowed to varying degrees with Kṛṣṇa's primordial potencies. Śrīla Prabhupāda gives the following description of some of the mystic *siddhis*:

> A mystic *yogi* can enter into the sun planet simply by using the rays of the sunshine. This perfection is called *laghimā*. Similarly, a *yogi* can touch the moon with his finger. Though the modern astronauts go to the moon with the help of spaceships, they undergo many difficulties, whereas a person with mystic perfection can extend his hand and touch the moon with his finger. This *siddhi* is called *prāpti*, or acquisition. With this *prāpti-siddhi*, not only can the perfect mystic *yogi* touch the moon planet, but he can extend his hand anywhere and take whatever he likes. He may be sitting thousands of miles away from a certain place, and if he likes he can take fruit from a garden there [NOD, pp. 11–12].

The *prāpti-siddhi* provides a perfect example of what we mean by the extension of access between locations. Consider the yogi on the earth who reaches out his hand to touch the moon. Does the yogi experience that his hand moves up through the atmosphere and crosses over thousands of miles of outer space, followed by a greatly elongated arm? This hardly seems plausible. We suggest that this *siddhi* actually allows the yogi to reach any desired location directly, and thus it requires higher-dimensional connections between remotely separated regions. The idea here is that Kṛṣṇa always has direct access to all locations, and by His grace this power of direct access can be conferred to varying degrees on various living beings.

The following verses in the Eleventh Canto of *Śrīmad-Bhāgavatam* (11.15.10–13) show that the *siddhis* are indeed obtained by partial realization of Kṛṣṇa's inherent potencies:

1. *aṇimā*—becoming smaller than the smallest. "One who worships Me [Kṛṣṇa] in My atomic form pervading all subtle elements [*bhūta-sūkṣma* and *tan-mātra*], fixing his mind on that alone, obtains the mystic perfection called *aṇimā*."

2. *mahimā*—becoming greater than the greatest. "One who absorbs his mind in the particular form of the *mahat-tattva* and thus meditates upon Me as the

Supreme Soul of the total material existence achieves the mystic perfection called *mahimā*."

3. *laghimā*—becoming lighter than the lightest. "I exist within everything, and I am therefore the essence of the atomic constituents of material elements. By attaching his mind to Me in this form, the yogi may achieve the perfection called *laghimā*, by which he realizes the subtle atomic substance of time."

4. *prāpti*—acquisition. "Fixing his mind completely on Me within the element of false ego generated from the mode of goodness, the yogi obtains the power of mystic acquisition, by which he becomes the proprietor of the senses of all living entities. He obtains such perfection because his mind is absorbed in Me."

Similar statements are made about the four other *siddhis*. According to the purport to SB 11.15.13, "Śrīla Bhaktisiddhānta Sarasvatī Ṭhākura states that those who pursue such perfections without fixing the mind on the Supreme Lord acquire a gross and inferior reflection of each mystic potency."

2.D. THE ACTIVITIES OF DEMIGODS, YOGIS, AND ṚṢIS

In the *Śrīmad-Bhāgavatam* there are many references to the mystic powers of demigods, yogis, and *ṛṣis*. These living beings are clearly endowed with more highly developed sensory powers than ordinary human beings, and they also are able to operate within a more extensive realm of activity than the space-time continuum of our ordinary experience. (Note that in accordance with Vedic usage, we are using the term "sensory" to refer both to senses of perception and to senses of action.)

A typical inhabitant of the higher planets has a life span of 10,000 celestial years, where each day and each night equals six earthly months (SB 4.9.63p). However, many demigods live for a much longer period. Thus demigods such as Indra hold official positions in the universal administration for the span of one *manvantara*, or 71 × 12,000 celestial years, and their total life span is much longer.

The demigods have the power to assume any desired form (SB 8.15.32p) and to appear and disappear at will before ordinary human beings. Thus SB 9.21.15 says that demigods such as Lord Brahmā and Lord Śiva appeared in human form before Mahārāja Rantideva, and SB 1.12.20p says that that Indra and Agni appeared before Mahārāja Śibi in the form of an eagle and a pigeon. There are also many passages in the *Bhāgavatam* that describe how demigods possessing higher levels of karmic merit can appear and disappear at will before lesser demigods. For example, Indra's guru, Bṛhaspati, made himself inaccessible to Indra after Indra offended him (SB 6.7.16).

Our thesis is that this ability to appear and disappear is not "just a matter of mystical power." Rather, it demonstrates an important feature of the physical world in which we live. This world contains many manifestations that are not accessible to us with our ordinary senses, but that are accessible to more highly developed beings, such as the demigods. There is a hierarchy of dimensional levels within the universe, and beings on one particular level can operate within a larger continuum than beings on lower levels. The spiritual realm of Vaikuṇṭha and Goloka Vṛndāvana is on a still higher level. Thus Brahmā, the topmost demigod within the material universe, became completely bewildered when Kṛṣṇa revealed the spiritual world to him.

In SB 1.16.3 it is said that during Mahārāja Parīkṣit's horse sacrifices, even a common man could see demigods. It appears that in Vedic times demigods often visited the earth and engaged in various dealings with human beings. Generally, however, only qualified persons were able to see them. Even recently, after the birth of Lord Caitanya, to glorify the Lord demigods used to visit the home of Jagannātha Miśra while remaining invisible (CC AL 14.76–81).

The *Bhāgavatam* often alludes to the idea that by acquiring higher spiritual qualifications one can enhance one's sensory powers and automatically experience phenomena within a broader realm of existence. (It is also emphasized, of course, that such powers should not be exploited for sense gratification, since this would divert one from the actual goal of spiritual life.) One example of such powers is indicated by Nārada Muni's instructing Dhruva Mahārāja that by chanting a certain mantra—*oṁ namo bhagavate vāsudevāya*—Dhruva would soon be able to see "the perfect human beings [*khe-carān*] flying in the sky" (SB 4.8.53).

One method that was sometimes used to travel between the higher planets and the earth is mentioned in SB 3.8.5p, where we read that great sages can travel from Satyaloka to the earth via the Ganges River, which flows all over the universe. Śrīla Prabhupāda points out that this form of travel is possible in any river by mystic power. It hardly seems plausible that this method of travel involves swimming up- or downstream over vast distances, and, of course, the connection between the earthly Ganges and its celestial counterpart is not visible to us. We suggest that this process of travel involves higher-dimensional connections between locations, and that the river serves as a kind of guiding beacon to direct such higher-dimensional transport. In the case of the Ganges, the course of the river from higher planets down to the earth must also be higher-dimensional.

In KB, p. 534 there is a description of the mystic *yoginī* Citralekhā traveling in outer space from Śoṇitapura to Dvārakā and taking the sleeping Aniruddha back to Śoṇitapura. This is another example of a form of travel that seems to require higher-dimensional connections for its operation.

The Vedic *śāstras* mention many remarkable events that are said to have taken place on the earth in the remote past. Many of these events involve phenomena that we do not experience today, and one might ask why this should be so, if these events actually did occur at one time. One reason for this given in the *Bhāgavatam* is that prior to the beginning of Kali-yuga, natural processes on the earth operated in a different mode than they do today (see SB 1.4.17–18p). The sensory powers of all living beings were on a higher average level than they are at present, and advanced beings such as demigods and great sages regularly visited the earth. Thus the earthly realm of ordinary human life was more intimately linked up with higher realms of material and spiritual reality than it has been since the start of the Kali-yuga.

This idea leads naturally to the following tentative scenario for the history of the last few thousand years: Once the Kali-yuga began, demigods and other higher beings greatly curtailed communications with people on the earth, and the general sensory level of human beings also declined. For some time, people continued to believe in stories about the earlier state of affairs on the earth due to the authority of tradition. However, due to the lack of feedback from higher sources and the natural cheating propensity of human beings, the traditions in various parts of the world gradually became more and more garbled, and people began to lose faith in them. Finally the present stage of civilization was reached, in which old traditions are widely viewed as useless mythology, and people seek knowledge entirely through the use of their current, limited senses.

2.E. REGIONS OF THIS EARTH NOT PERCEIVABLE BY OUR SENSES

We have been developing the idea that the three-dimensional continuum of our experience does not constitute the totality of spiritual or material reality. One feature of this idea is that there exist worlds, or realms of experience, that are located here on the earth but that cannot be perceived or visited by human beings possessing ordinary sensory powers. Of course, the most striking example of this is Kṛṣṇa's transcendental *dhāma* of Vṛndāvana. In CC Adi 5.18 purport it is stated that although Kṛṣṇa's abode is unlimited and all-pervading, it is identical to the Vṛndāvana of this earth. This implies that within the tract of land called Vṛndāvana in India, there exists a completely real domain of spiritual existence that is not accessible to the senses of ordinary conditioned beings. This is another example of higher-dimensional connections, and it implies that two (or more) worlds of experience can co-exist in parallel, in the same location.

The holy *dhāma* of Navadvīpa is another example of this (and, of course, Navadvīpa *dhāma* is also identical to Vṛndāvana). Śrīla Bhaktivinoda Ṭhākura says in the *Navadvīpa Mahātmyā,* "The *dhāma* of Navadvīpa, within Gaura Maṇḍala and served by the Gaṅgā, is situated in eternal splendor. . . . The form of Gaura Maṇḍala, eternally transcendental to the material world, is like the sun. The materialist's eye is covered by the cloud of illusion, and because of this he sees only the secondary transformations of that spiritual energy, the dull, inert material world" (NM, p. 4).

The transcendental realms of Navadvīpa and Vṛndāvana are purely spiritual, but there are also material examples illustrating the idea of parallel worlds co-existing in one place. For example, the *Bhāgavatam* states that Maru and Devāpi, two ancient royal princes belonging to the Sūrya and Soma dynasties, are still living in the Himalayas in a place called Kalāpa-grāma. By the power of mystic yoga they will prolong their lives until the beginning of the next Satya-yuga and then revive the lost Sūrya and Soma dynasties by begetting children (SB 9.12.6, 9.22.17–18, and 12.2.37–38).

If we go to the Himalayas, we will certainly not be able to perceive Maru and Devāpi using our ordinary senses, even though they are human beings possessing gross material bodies. It can also be argued that we will not be able to perceive the surroundings in which they live. A human being cannot live without interacting with his material surroundings. Even a yogi who is simply living on air requires an undisturbed sitting place. Could it be that the material accoutrements and sitting places of these two persons are directly visible and accessible to us, even though they themselves are invisible? We suggest that they are actually living in a setting that is entirely inaccessible to our senses, but that can be seen and entered by a person, such as an advanced yogi, whose senses can operate on an appropriate level.

Here the objection may be raised that a co-existing invisible world cannot be on the same level of reality as our world because it must be "subtle," transparent, or ghostlike in nature, whereas our own world is opaque and substantial. Our reply is that such a co-existing world is invisible to us not because it is made of transparent substance distributed within our own three-dimensional continuum, but rather because it lies in a higher dimension and is entirely outside our continuum. It can be in the "same place" as we are by virtue of higher-dimensional interconnection. A person with higher sensory powers is able to perceive this world not because he can discern some nearly transparent substance lying within his own three-dimensional space, but because his senses are not restricted to three dimensions and have access to broader realms of material or spiritual reality.

We should note that the basic elements—of earth, water, fire, air, and ether—are present in some form on all levels of reality, both spiritual and mundane. In SB 11.21.5 it is stated that these five elements constitute the bodies of all conditioned souls, from Lord Brahmā down to the nonmoving creatures. Also, CC AL 5.53 states that "the earth, water, fire, air, and ether of Vaikuṇṭha are all spiritual. Material elements are not found there."

The five material elements (*pañca-bhūta*) are described in the *Bhagavad-gītā* as separated energies of Kṛṣṇa. Their counterparts in Vaikuṇṭha are evidently similar enough to them to warrant being called by the same names. However, the spiritual elements must belong to Kṛṣṇa's internal potency. It would therefore seem that the spiritual world and the material world are similar in the sense that both contain variegated forms composed of solid, liquid, and gaseous constituents. At the same time, they have distinct qualitative features, of which one of the most notable is the presence of the modes of passion and ignorance in the material world and their absence in the spiritual world. Material realms on various dimensional levels will also possess similar variegated forms, but the higher realms will be characterized by greater predominance of the mode of goodness over the modes of passion and ignorance.

As a final point, we note that the history of the Mādhva-Gauḍīya-sampradāya sheds some light on the higher-dimensional nature of reality. In SB 1.4.15p Śrīla Prabhupāda points out that Vyāsadeva is residing in Śamyāprāsa in Badarikāśrama. Many people in India make a pilgrimage to Badarikāśrama every year, but it is not possible for an ordinary person to meet Vyāsadeva. However, it is said that Madhvācārya met Vyāsadeva there and took initiation from him. It was through this higher-dimensional link that the Mādhva-Gauḍīya-sampradāya was passed down from Śrīla Vyāsadeva to the recent line of acharyas.

Vedic Cosmography

Śukadeva Gosvāmī said: "My dear King, there is no limit to the expansion of the Supreme Personality of Godhead's material energy. This material world is a transformation of the material qualities. . . , yet no one could possibly explain it perfectly, even in a lifetime as long as that of Brahmā" (SB 5.16.4).

In this chapter we will describe the structure of this universe, or *brahmāṇḍa*, as described in the *Śrīmad-Bhāgavatam*. Our aim is to show the relation between the Vedic picture of the universe and the world of our experience. In doing this, we will draw information from the following sources: (1) The writings of Śrīla Prabhupāda, including his translation, with commentary, of the *Śrīmad-Bhāgavatam*, (2) other writings in the Vedic tradition, including the *Sūrya-siddhānta*, and (3) modern Western science.

Our strategy is to present the simplest possible world-picture that will harmonize (1) and (3), given the assumption that the *Bhāgavatam* gives a direct and valid account of the universe. In doing this, we will try as far as possible to avoid introducing speculative hypotheses. (This is Newton's principle of *hypothesis non fingo*.) This means that we will often have to make statements of the form "A corresponds to B," without spelling out the exact nature of the correspondence. In some cases we will try to show the plausibility of the correspondence by offering a speculative explanation of how it might come about. However, all explanations of this kind should be regarded as tentative and subject to correction in the future.

3.A. BHŪ-MAṆḌALA, OR MIDDLE EARTH

The Vedic literature describes the material cosmos as an unlimited ocean situated within a small part of the unlimited spiritual world. Within this ocean there are innumerable universes, or *brahmāṇḍas*, which can be compared to spherical bubbles of foam grouped in clusters. Each of these universal globes consists of a series of spherical coverings and an inner, inhabited portion.

Within the inner region of the *brahmāṇḍa*, the most striking feature is Bhū-maṇḍala, or the earthly planetary system. Bhū-maṇḍala is described in the Fifth Canto of *Śrīmad-Bhāgavatam* as a flat disc with a diameter of 500 million *yojanas*, or 4 billion miles (using 8 miles per *yojana*). The surface of this disc is marked with a series of ring-shaped oceans and islands surrounding a central island called Jambūdvīpa.

The total surface area of our familiar earth planet is some 197 million square miles, and, according to modern information, the total surface area of the sun is about 2.4 million million square miles. In contrast, the total area of Bhū-maṇḍala comes to about 12.6 billion billion square miles. In SB 2.5.40p Śrīla Prabhupāda refers to this as the area of the universe, and it seems that Bhū-maṇḍala is indeed one of the most significant and frequently mentioned features in the Vedic account of the universe. Its size is on the scale of the solar system as a whole, as conceived in modern Western astronomy.

The Fifth Canto gives specific figures for the size, shape, and position of many of the geographic structures of Bhū-maṇḍala. The most striking characteristic of these structures is that although their description employs names for familiar features of earthly geography, such as mountains, oceans, and islands, they are all on the same cosmic scale as Bhū-maṇḍala itself. Thus the smallest mountains on Bhū-maṇḍala mentioned in the *Bhāgavatam* are 2,000 *yojanas*, or 16,000 miles, high. Many mountains are 80,000 miles or even 672,000 miles high. In contrast, the diameter of the earth is about 8,000 miles, and Mount Everest, the highest known mountain, extends about 5.5 miles above sea level. References to such immense sizes are not limited to the Fifth Canto. For example, SB 4.6.32 gives a description of Lord Śiva meditating underneath a banyan tree 800 miles in height and 600 miles in breadth. In SB 8.2.1 we read that Trikuta Mountain, where the elephant Gajendra achieved liberation, is 80,000 miles in length and breadth. This mountain is situated in the ocean of milk, one of the geographical features of Bhū-maṇḍala. In SB 8.7.9 it is pointed out that when Kūrma, the tortoise incarnation of Lord Viṣṇu, was supporting Mandara Mountain during the churning of the milk ocean, His back extended for 800,000 miles (*lakṣa-yojana*), "like a large island." Finally, the Matsya *avatāra*, Lord Viṣṇu's fish incarnation, expanded from an initial small size to a final length of 8 million miles (SB 8.24.44).

Modern scholars tend to reject dimensions such as these as ludicrous exaggerations made by poets who were completely devoid of scientific knowledge. However, even common men in primitive societies can tell that the earthly mountains of our experience have heights of thousands of feet rather than thousands of miles. The highly rational philosophical discussions in the *Bhāgavatam* indicate that it was not written by some kind of mad fanatic who was devoid of common sense. We suggest, therefore, that the descriptions in the *Bhāgavatam* of gigantic sizes refer to an actually existing world that is built on the scale of the solar system and that contains features built on a similar scale.

We will assume that this is the case, and later on we will consider what the relation might be between this world and the earth of our experience. For the present we will give a brief overview of the most significant features of Bhū-maṇḍala. We will do this with the aid of a series of computer-generated illustrations that portray the features of Bhū-maṇḍala as they would appear to an observer approaching Bhū-maṇḍala from a great distance.

In the first view (Figure 3) we are looking down on the center of Bhū-maṇḍala at an angle of 45° from a distance of some 600 million miles. We can discern five ring-shaped structures surrounding a central region that is too far away to see clearly. Going from the outside in, these are respectively the *dvīpas*, or islands, named Puṣkaradvīpa, Śākadvīpa, Krauñcadvīpa, Kuśadvīpa, and Śālmalīdvīpa. Puṣkaradvīpa has inner and outer radii of 100.4 million and 151.6 million miles, and each successive ring, going inward, is half as wide as the one preceding it. To give an idea of the scale, the distance from the earth to the sun is currently accepted to be 93 million miles.

Figure 3. The plane of Bhū-maṇḍala, looking down at an angle of 45° and at a distance of 600 million miles from the center. (This is the apparent distance if the figure is viewed from about two feet.) The *dvīpas* named Puṣkaradvīpa, Śākadvīpa, Krauñcadvīpa, Kuśadvīpa, and Śālmalīdvīpa, appear as concentric rings, going toward the center from the outside. Plakṣadvīpa and Jambūdvīpa appear to be merged together at the center.

Figure 4. The plane of Bhū-maṇḍala, seen at a distance of 150 million miles from the center. Now Jambūdvīpa is barely visisble in the center of the innermost ring, called Plakṣadvīpa.

The intervals between the *dvīpas* are occupied by oceans, each of which has the same width as the *dvīpa* that it surrounds. The oceans surrounding the five *dvīpas* we have mentioned are said to be composed respectively of clear water, yogurt, milk, ghee, and liquor. Of course, these substances are celestial counterparts of the corresponding ordinary substances of our day-to-day experience.

In Figure 4 we have moved in to a distance of about 150 million miles from the center of Bhū-maṇḍala. Now Krauñcadvīpa, Kuśadvīpa, and Śālmalīdvīpa have expanded in apparent size, and the ring of Plakṣadvīpa has become visible within Śālmalīdvīpa. We can also begin to discern the central island of Jambūdvīpa within Plakṣadvīpa. In Figure 5 we have moved in to a distance of 15 million miles, and in Figure 6, at a distance of some 3 million miles, we can obtain a detailed view of Jambūdvīpa.

Jambūdvīpa is described as a disc-shaped island 100,000 *yojanas*, or 800,000 miles, in diameter. (For comparison, the currently accepted diameter of the sun is 865,110 miles.) The most striking feature of Jambūdvīpa is a central structure called Mount Meru, which is 84,000 *yojanas* high. This structure is generally referred to as a mountain, although it clearly has a unique form quite different from that of a typical mountain. The upper surface of Mount Meru is said to be occupied by Brahmapurī, the city of Lord Brahmā, and by cities belonging to eight other demigods.

Jambūdvīpa is divided into nine regions, or *varṣas*, by a series of mountain ranges. In Figures 7 and 8 we see a more detailed view of the central region of Ilāvṛta-varṣa, which contains Mount Meru and is square in shape. To get some idea of the scale of this figure, we should note that the low mountain chain stretching from A to B in the Figure 7 is called Gandhamādana and reaches 16,000 miles in height. This is twice the diameter of the earth.

Of the nine *varṣas* of Jambūdvīpa, eight are described as places of heavenly enjoyment. These are intended for persons who have returned to earth after using up their allotted time on the heavenly planets but who have some remaining pious credits entitling them to enjoy great material opulence. The inhabitants of these *varṣas* are described as living for 10,000 years by earthly calculations, as having the bodily strength of 10,000 elephants, and as having a standard of pleasure like that of the human beings of Tretā-yuga (SB 5.17.12). These regions are also said to contain beautiful gardens that are visited by important leaders among the demigods. The *Bhāgavatam* refers to these eight *varṣas* as *bhauma-svarga*, or the heavenly places on earth (SB 5.17.11), while Śrīla Prabhupāda describes them as "the lower heavenly planets" and contrasts their inhabitants to those of "this earth" (SB 5.17.13p).

Figure 5. The plane of Bhū-maṇḍala, seen from a distance of 15 million miles. Jambūdvīpa and the surrounding salt-water ocean are now visible.

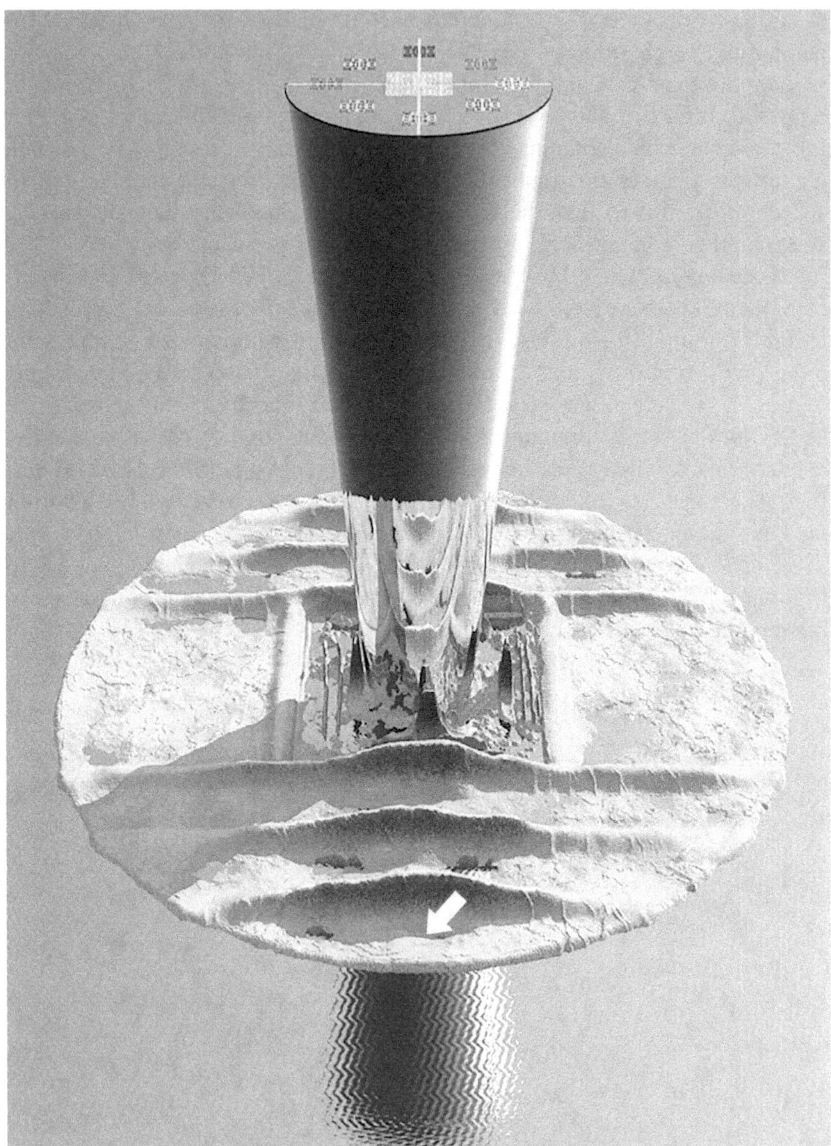

Figure 6. A detailed view of Jambūdvīpa from the south. The inverted cone is Mount Sumeru, and on its summit lies Brahmapurī and the residents of eight principal demigods. The smaller mountain ranges form the boundaries of the nine *varṣas*. The arrow indicates the location of Bhārata-varṣa. This picture is rendered according to the descriptions given in the Fifth Canto of the *Śrīmad-Bhāgavatam*.

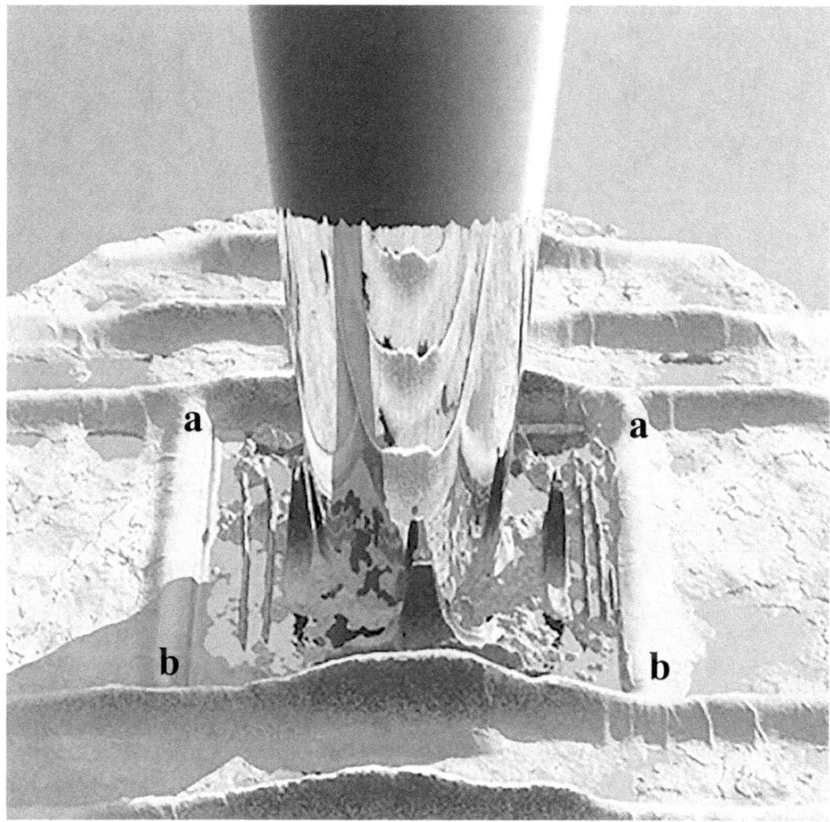

Figure 7. A closer view of Ilāvṛta-varṣa, the central subdivision of Jambūdvīpa. Only those mountains are shown for which dimensions are given in the *Śrīmad-Bhāgavatam*. The mountain ranges stretching from **a** to **b** are 16,000 miles in average height, or twice the diameter of the earth.

The remaining *varṣa* of Jambūdvīpa is called Bhārata-varṣa. It is described as the field of fruitive activities, in which human beings struggle with adverse conditions and elevate or degrade themselves by their actions. Bhārata-varṣa is the southernmost region of Jambūdvīpa, and its location is indicated by arrows in Figures 6 and 9. In shape, Bhārata-varṣa is a semicircular piece of land bounded on the south by the salt-water ocean and on the north by the Himalayan Mountains. Bhārata-varṣa is the only part of Bhū-maṇḍala at all reminiscent of the earth, and it is frequently identified with either the earth or with India in Śrīla Prabhupāda's books.

Yet the Himalayas bounding Bhārata-varṣa are described in the Fifth Canto

Figure 8. Another view of Ilāvṛta-varṣa, with mountains reflecting from the base of Mount Sumeru.

as being 80,000 miles high, and Bhārata-varṣa itself runs some 72,000 miles (9,000 *yojanas*) from north to south. This naturally leads us to ask, What is the relationship between the earth of our experience and Jambūdvīpa and Bhārata-varṣa, as described in the *Śrīmad-Bhāgavatam*?

3.B. THE EARTH OF OUR EXPERIENCE

In this book we will take it for granted that the earth planet on which we live our daily lives can be practically thought of as a globe with a diameter of about 8,000 miles. In the age of international travel by jet airplanes, it is easy for people in general to accumulate abundant evidence that confirms this. Commercial airlines fly regularly scheduled flights along a network of routes that completely covers the inhabited areas of the earth. A glance at an airline's route map shows that each of these routes follows a great circle—the shortest path connecting two points on the surface of a sphere. (There are some exceptions, of course, due to political considerations.) One can experience changes in time zones of the kind that one would expect to find if the earth is a globe, and one can consider that if the airline authorities do not properly understand the size and shape of the earth, along with the location of various cities on it, then how is it possible for them to arrange regular flights from one city to another?

There are many regions on the earth that have not been thoroughly explored. However, it would be difficult to argue that airplanes have not flown over most areas of the earth's surface, including the Arctic and Antarctic regions. One can read popular articles describing life during the winter at an American base at the South Pole, and one can also read about artificial satellites with orbits ranging from equatorial to circumpolar. Thus human experience with remote, seldom-visited regions of the earth is also consistent with the idea that the earth is a sphere.

Yet, even though the earth can be regarded as a globe from the viewpoint of our ordinary sensory experience, we have already argued that there is a sense in which the earth is definitely not a globe. The very idea of a sphere is based on three-dimensional Euclidian geometry. Thus, if the three-dimensional continuum of our ordinary experience is simply a limited aspect of a higher-dimensional reality, it follows that the globe of the earth is also simply an aspect of that higher reality. To properly describe what that reality is, in and of itself, we must go beyond three-dimensional constructs such as a sphere or a plane. A yogi who can reach directly to another continent by means of the *prāpti-siddhi* is not experiencing the earth as a sphere. Similarly, a person who

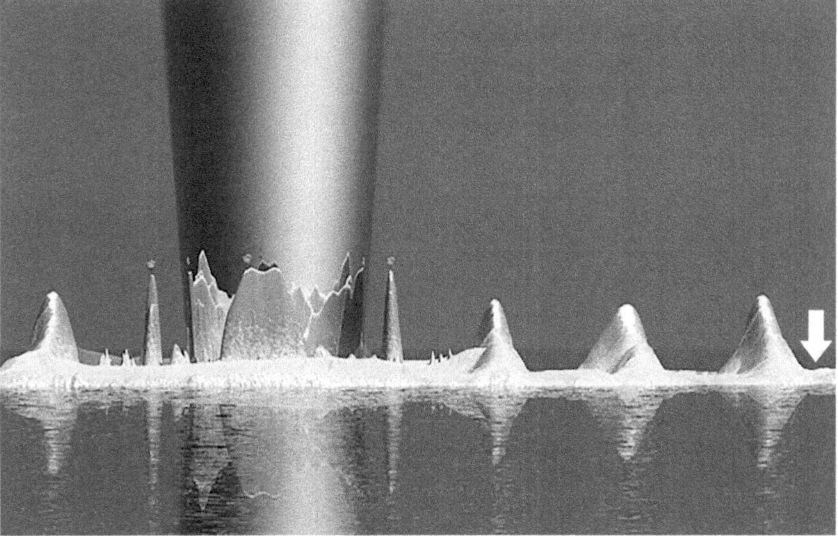

Figure 9. Jambūdvīpa as seen from the west at a lower elevation. The arrow points to Bhārata-varṣa, located between the base of the Himalayan Mountains and the continental edge.

is able to realize that Vṛndāvana in India is nondifferent from the unlimited spiritual realm of Goloka cannot be thinking of the earth simply as a small globe. The earth globe may be one aspect of the reality that he is experiencing, but he may choose to describe that reality by emphasizing other aspects that for him are more important.

We propose that although the total reality of the world is very difficult, or even impossible, to fully describe in words, different aspects of it can be described in readily comprehensible language. These aspects correspond to different perspectives, which depend on the different situations and sensory capacities of different observers. Simple geometric imagery may be quite fitting for the description of the universe from many of these different individual perspectives, even though it is completely inadequate to describe the material world as a whole.

In this book we propose that the cosmological system of the *Bhāgavatam* is a simplified description of the universe as it appears from the viewpoint of demigods, *ṛṣis*, and highly elevated human beings, who are the principal characters in this work. In contrast, our familiar conception of the earth globe is a valid account of our immediate environs as they appear from the viewpoint of persons with ordinary human senses. This can also be said of the world system of the astronomical *siddhāntas*, which we have proposed in Chapter 1 to be an integral and long-standing part of the Vedic culture. There the earth is also described as a small globe, and the astronomical discussions are limited to phenomena that people can observe with their gross senses.

3.b.1. Bhārata-varṣa

In an abstract form, the foregoing is our general idea about the nature of the relationship between Vedic cosmology and our modern world view. However, to make this idea more vivid and concrete, it is necessary to work it out in much greater detail. We will now proceed to do this, beginning with the question of how this earth relates to Bhū-maṇḍala as a whole.

In SB 5.19.21p Śrīla Prabhupāda refers to Bhārata-varṣa as India, and he points out that the demigods aspire to take birth there. In SB 2.7.10p this earth planet is identified with Bhārata-varṣa, and a similar reference is made in SB 1.12.20p. In SB 3.18.19p Śrīla Prabhupāda points out that the earth planet was once known as Ilāvṛta-varṣa, but when Mahārāja Parīkṣit ruled the earth it was called Bhārata-varṣa. By the process of political fragmentation, Bhārata-varṣa gradually came to mean India alone. The idea that Bhārata-varṣa once referred to the entire earth is also indicated in SB 4.22.36p, where Śrīla Prabhupāda suggests on the basis of Purāṇic references that Brazil, rather than Ceylon, was Rāvaṇa's kingdom.

In SB 1.12.5 the earth ruled by Mahārāja Yudhiṣṭhira is referred to as Jambūdvīpa, and in SB 4.12.16 the earth ruled by Dhruva Mahārāja is referred to as Bhū-maṇḍala itself. Going further, SB 5.1.22 states that Mahārāja Priyavrata ruled all the planets of the universe (*akhila-dharā-maṇḍala*), and Śrīla Prabhupāda points out that it is difficult for us to understand just where Mahārāja Priyavrata was situated.

In addition to his statements identifying our earth with Bhārata-varṣa, Śrīla Prabhupāda also makes statements indicating that some regions of Bhū-maṇḍala are not part of this earth. We have already noted his reference to the other eight *varṣas* of Jambūdvīpa as "the lower heavenly planets." In SB 4.18.20 one of these *varṣas*, known as Kiṁpuruṣa-varṣa, is spoken of as a planet whose inhabitants are endowed with remarkable mystic powers. In SB 3.23.39p Śrīla Prabhupāda describes Mount Meru as being a resort area for demigods that is "situated somewhere between the sun and the earth," and in SB 3.2.8p he says that the moon was born from the milk ocean "in the upper planets." In SB 5.1.8p he speaks of "a planet covered mostly by great mountains, one of which is Gandhamādana Hill." This mountain marks one of the boundaries of Ilāvṛta-varṣa (SB 5.16.10). When commenting on the description of Ilāvṛta-varṣa in SB 5.16.10, he distinguishes between the mountains of this planet earth and the "greater mountainous areas of the universe." Finally, in SB 8.2.14–19p he describes Trikūṭa Mountain, surrounded by the ocean of milk, as being on another planet.

All of these statements can be reconciled if we adopt the idea that the earth of the *Bhāgavatam* is the disc of Bhū-maṇḍala, but that only a small portion of this earth is accessible to the limited senses of modern-day human beings. In previous *yugas* larger regions of Bhū-maṇḍala were accessible, and people experienced a correspondingly larger earth. Thus in Mahārāja Yudhiṣṭhira's time, at the end of the Dvāpara-yuga, people had access to the entire region of Jambūdvīpa, and people living in the Satya-yuga during the reign of Dhruva Mahārāja had access to the whole of Bhū-maṇḍala. In the *Caitanya-caritāmṛta* it is said that persons from the various *dvīpas* of Bhū-maṇḍala visited the home of Lord Caitanya disguised as human beings. To these persons it is presumably still natural to think of the earth as Bhū-maṇḍala.

3.b.2. The Projection of Bhū-maṇḍala on the Sky

If Bhū-maṇḍala is a disc 4 billion miles in diameter, one natural question is, Where is this disc located? We have indicated that our own location on the earth corresponds to part of Bhārata-varṣa, which lies almost exactly in the center of Bhū-maṇḍala. In SB 1.1.4p we read that Lord Brahmā once envisioned the forest of Naimiṣāraṇya in India as the center of a great wheel that enclosed the universe. This suggests that this well-known site in India is located exactly in the center of the vast disc depicted in Figure 3. In any case, both India and the rest of the earth of our experience must lie close to this center.

Let us consider a person somewhere on this earth. If he is standing in the center of a disc that extends for millions of miles into space, then from his perspective most of that disc will be very far away from him, and it will appear to be projected into a circular band running through the heavens. We can discuss this circular band more precisely by introducing the celestial sphere of the astronomers.

In observational astronomy it is customary to visualize celestial objects such as stars and planets as lying on the surface of an enormous imaginary sphere

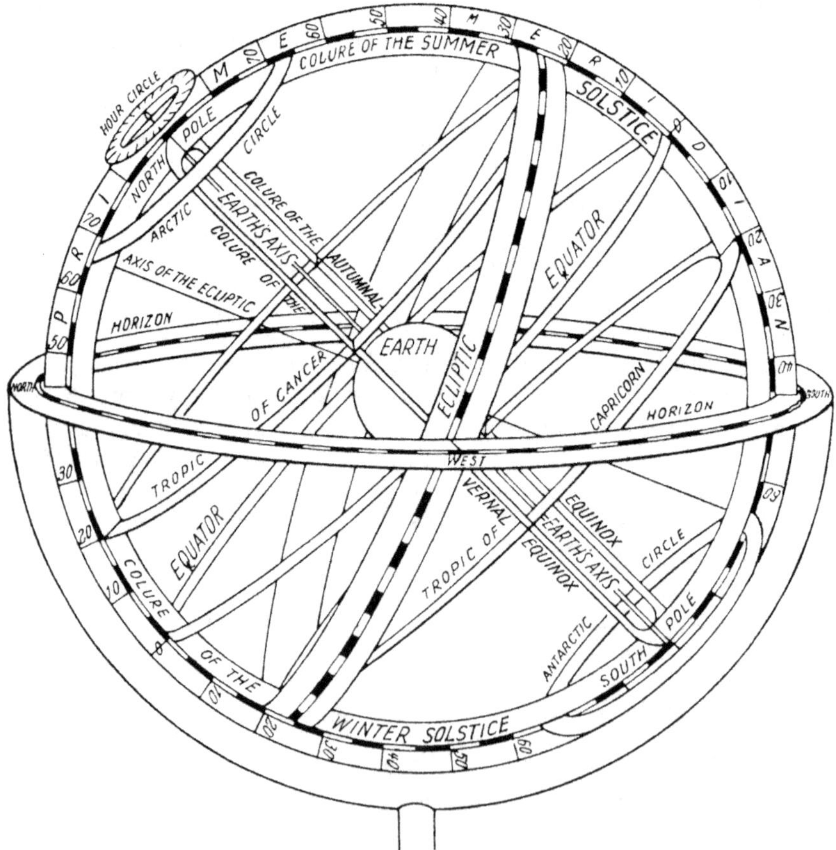

Figure 10. The celestial sphere, showing the ecliptic and the celestial equator.

centered on the earth. The system of earthly latitude and longitude is projected onto this sphere, and thus the sphere has celestial north and south poles corresponding to the north and south poles of the earth, and also a celestial equator corresponding to the earth's equator (see Fig. 10). Any disc centered on the earth and extending millions of miles into space will intersect this sphere in a great circle tilted at some angle to the celestial equator. Our question thus becomes, What great circle on the celestial sphere corresponds to the disc of Bhū-maṇḍala?

In Section 3.d we will discuss the daily and yearly motion of the sun. We will argue that the projected orbit of the sun on the celestial sphere provides a marker that will help us locate the projection of Bhū-maṇḍala. We will present two hypotheses regarding this projection: (1) The projection of Bhū-maṇḍala coincides with the great circle known as the ecliptic. This circle marks the yearly path of the sun through the heavens, and it passes through the circular band of constellations known as the zodiac. (2) The projection of Bhū-maṇḍala corresponds to the celestial equator. Although we tend to favor hypothesis (1), we present both hypotheses, since some śāstric support can be provided for each one.

We will discuss these hypotheses in detail in Section 3.d, but for the moment we will consider some questions that naturally arise from them. First of all, it might seem that the *Bhāgavatam* is presenting a simple model of the earth as a flat plane. According to this idea, the plane of Bhū-maṇḍala should be parallel to the surface of the earth, and therefore the projection of Bhū-maṇḍala on the sky should correspond to the circle of the horizon. One problem with this is that the *Bhāgavatam* contains a number of verses indicating that the sun moves in a circular path on the surface of Bhū-maṇḍala (or very close to it) at a very large distance from Jambūdvīpa (for example, see SB 5.20.30). If the celestial projection of Bhū-maṇḍala corresponds to the horizon, then these verses imply that the sun must always remain close to the horizon, instead of rising in the east, going high in the sky, and setting in the west as we observe. In fact, the Indian astronomer Bhāskarācārya seems to believe that the *Purāṇas* do imply this, and he takes this as a reason for rejecting the Purāṇic view (SSB1, p. 114).

Actually, in the Arctic and Antarctic regions the sun does behave in this way at certain times of the year. However, since the earth of our experience is a globe, the inclination of the sun's path in the sky changes as we go north and south, and over most of the earth's inhabited regions this inclination is very steep. In Chapter 1 we have argued that the spherical nature of this earth planet was known in Vedic times, and this, of course, is incompatible with a flat-earth interpretation of Vedic cosmology. However, even if we disregard this point, we can hardly suppose that a hypothetical pre-scientific sage living by the side of the Ganges would not have noticed that the sun moves high overhead in the course of a day. We therefore propose that the *Purāṇas* could not be identifying

Figure 11. The stars visible to the naked eye in the northern celestial hemisphere.

the plane of Bhū-maṇḍala with the horizon.

At this point, the objection will be raised that when we look at the sky at night, we do not see anything unusual in the direction of either the zodiac or the celestial equator. Indeed, we see nothing but stars in all directions. If the surface of Bhū-maṇḍala bisects the sky along one of these great circles, then we should see stars only on one side of the circle. On the other side we should see solid earth, as we do in the case of the horizon. Our answer to this objection is that since most of Bhū-maṇḍala is not accessible to our senses, we cannot see it.

This may initially seem to be a rather unsatisfactory answer, but it is consistent with all of the material that we have gathered from the *Bhāgavatam* thus far. For example, the height of Mount Meru is nearly equal to the diameter

Figure 12. The stars visible to the naked eye in the southern celestial hemisphere.

of the sun (according to modern data), so if it is indeed located "somewhere between the sun and the earth," then why can't we see it? Also, if the plane of Bhū-maṇḍala exists at all, and acts as a barrier to our vision, then the sky must be bisected along *some* circle, with all visible stars lying on one side. Yet, if we go from the Northern to the Southern Hemisphere, it is possible to look at the night sky in all directions, and wherever we look, we simply see stars. This is true, for example, if we look towards the south celestial pole from New Zealand or South Africa (see Figs. 11 and 12).

Another question that may be raised is, If you are saying that Bhū-maṇḍala is higher-dimensional and therefore invisible, why do you try to assign it a location in three-dimensional space at all? The answer is that a higher-dimensional structure can also have a three-dimensional location. To illustrate this idea, consider a person who is trying to find a particular office in Manhattan. By moving north-south and east-west through the grid of streets, he may arrive at the address of the office but be disappointed to find that he cannot see it. To actually reach the office he may have to move fifty stories in the vertical direction by taking an elevator. Thus, the office has a two-dimensional location, but to reach it, three-dimensional travel is necessary. Likewise, to reach a given location in Bhū-maṇḍala, both three-dimensional and higher-dimensional travel may be required.

In summary, we propose that the Vedic cosmology corresponds to our observable world in the following way: The earth of our experience is a small globe surrounded by the starry heavens in all directions. Bhū-maṇḍala is a vast disc that extends for millions of miles into space but is not perceivable by our present senses. Its projection on the celestial sphere must be ascertained on the basis of the movement of the sun, and this projection does not correspond to the variable horizon of this earth. We suggest that this is not simply an artificial reconciliation of Vedic cosmology with modern astronomical views. Rather, we propose that this is how Vedic cosmology was understood in ancient times.

3.b.3. A Historical Interlude

In this subsection we will briefly consider some historical evidence suggesting that Vedic cosmology, or something very similar to it, may once have been widely accepted throughout the world. Some of this evidence supports the ideas we have just outlined on the nature and position of Bhū-maṇḍala.

Societies throughout the world have traditionally passed down ancient legends and myths describing the nature and origin of the universe. In this seemingly chaotic array of diverse stories, two historians named Giorgio de Santillana and Hertha von Dechend thought they could see evidence for a common ancestral culture. According to them, this "archaic" culture antedated all the ancient civilizations we know of today, including those of Babylon, China, and India. They argued that this culture possessed a sophisticated scientific understanding of astronomy, but that it expressed this understanding in terms we today call mythological because we do not understand them.

Here is what de Santillana and von Dechend have to say about how this archaic culture viewed the earth:

> (1) First, what was the "earth"? In the most general sense, the "earth" was the ideal plane laid through the ecliptic. The "dry earth," in a more specific sense, was the ideal plane going through the celestial equator [HM, p. 58].
>
> (2) The name of "true earth" (or of the "inhabited world") did not in any way

> denote our physical geoid for the archaics. It applies to the band of the zodiac, two dozen degrees right and left of the ecliptic [HM, pp. 61–62].
> (3) At the "top," in the center high above the "dry" plane of the equator, was the Pole star. At the opposite top, or rather in the depth of the waters below, unobserved from our latitudes, was the southern pole, thought to be "Canopus" [HM, p. 63].

The idea of the earth presented here runs parallel to the ideas we have discussed regarding Bhū-maṇḍala. According to the *Bhāgavatam*, below the plane of Bhū-maṇḍala are seven lower planetary systems and then the Garbhodaka Ocean, which fills one half of the universal globe. Here we see a similar conception of the earth as a plane projected against either the celestial equator or the band of the zodiac, with a region of water in the direction of the southern pole.

Many bits and pieces of information can be collected from old myths and legends suggesting that a cosmology similar to that of the *Bhāgavatam* was widely disseminated in ancient times. In many cases this information comes down to us in the form of what may be called fossilized stories, or stories that have lost their original meaning but have been preserved in a distorted, fragmentary form in various traditions. One interesting example of this is the following story taken from Norse mythology: At the time of the destruction of the cosmos (the Norse *ragnarok*), all-engulfing flames come out of Surt the Black. This Surt is said to be "the king of eternal bliss 'at the southern end of the sky.'" It is also stated in the Norse myths that "there are many good abodes and many bad; best it is to be in Gimle with Surt" (HM, p. 157).

Here one cannot help but think of Saṅkarṣaṇa, or Ananta Śeṣa, who destroys the three worlds with fire at the time of annihilation, and who reclines on the Garbhodaka Ocean. If we project the location of Saṅkarṣaṇa on the sky, it should be to the south, in the direction of the watery region mentioned in (3) above. Saṅkarṣaṇa is known as *tāmasī*, or "dark," since He is in charge of annihilation, but He is also certainly the king of eternal bliss (SB 5.25.1). This passage from Norse mythology is therefore very curious, since standard historical accounts describe the ancient Scandinavians as polytheists who had no conception of one Supreme Godhead.

Whatever the true significance of the story of Surt may be, the ancient Scandinavians clearly had a concept of the earth that is very similar to Jambū-dvīpa as described in the *Bhāgavatam*. They regarded the earth as a circular island surrounded by a world ocean. In the center of the island is an enormous mountain, crowned by Asgard, the home of the gods (see Fig. 13). Interestingly enough, the number of warriors of the gods stationed in Asgard is 432,000, a number that often appears in the Vedic literature (HM, p. 162).

In his book *Shamanism* (SH), Mircea Eliade points out that the idea of three worlds with a universal axis marked by a cosmic mountain is extremely widespread. It is found in the ancient cultures of Egypt, India, China, Greece, and Mesopotamia, and it is also found in tribal societies throughout Asia, Africa, and the Americas. In central Asia, the names for the central mountain, such as Sumber, Sumur, or Sumer, are clearly related to the Sanskrit name Sumeru. The Greeks, of course, had their Mount Olympus; the Iranians had Haraberezaiti (Elbruz); the Germans had Himingbjorg; the Saxons had Irminsul, "the universal column that sustains everything" (SH, p. 261); and the Chinese had Mount Khun-Lun, where the dwellings of the immortals were situated (ND, pp. 566–67). Among the Babylonians, the ziggurat represented the cosmic mountain, and the central pillar of tribal dwellings in Asia and North America

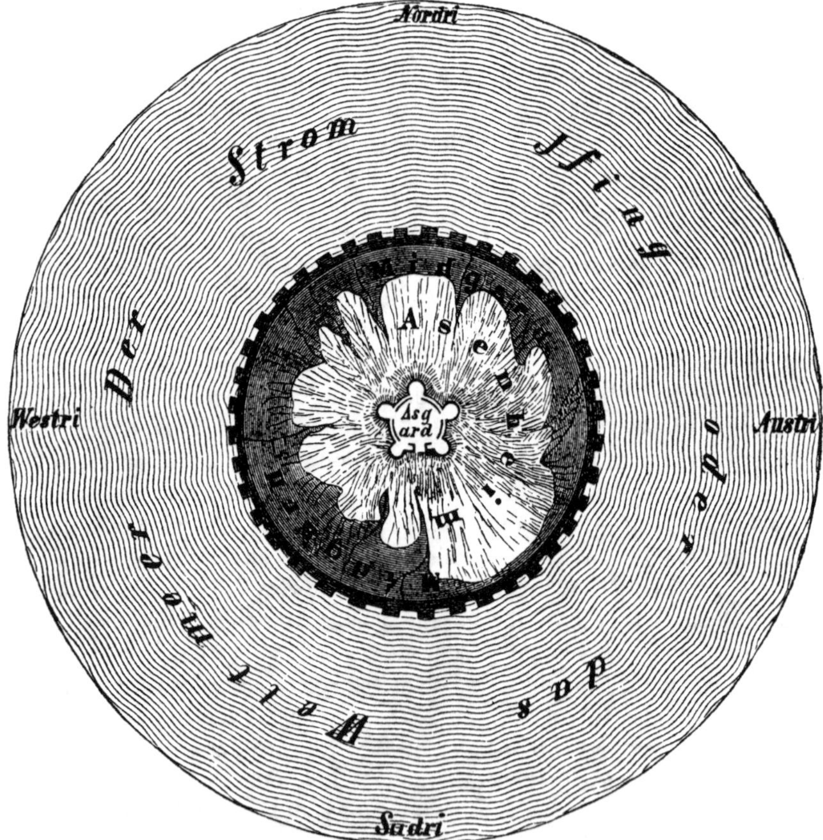

Figure 13. The Norse conception of the earth as a circular island surrounded by a world ocean. Asgard, the abode of the gods, is situated on the summit of an enormous mountain in the center of the earth.

carried a similar symbolic meaning (SH, pp. 261–62). Needham, in his *Science and Civilization in China*, notes that "wheel maps," depicting the earth as a circle surrounding a central mountain, were very common in the ancient world. He is uncertain as to whether these maps had ultimately an Indian or a Babylonian origin, but he notes that they seem to represent a tradition of great antiquity in both places (ND, pp. 588–90).

It may perhaps seem far-fetched to link the traditions of North American Indians with Vedic civilization, but even here we find some suggestive connections. For example, the Sioux Indians tell of a cycle of four ages. There is a buffalo that loses one leg during each age; at present we are in the last age—an age of degradation—and the buffalo has one leg. In the *Bhāgavatam*, of course, the same story is told about the bull named Dharma; at present we are in the last age (the Age of Kali), and Dharma is standing on one leg (EB, p. 9).

In Figure 14 we give another example of what may be a remnant of the Vedic world view. This is a picture from the *Maya Codex Tro-Cortesianus*. Some people have interpreted it as a depiction of the churning of the milk ocean, as described in the Vedic literature. The picture is difficult to interpret, but it does seem to contain a tortoise, a central churn, and a serpent being pulled like a rope by what may be demigods and *asuras*. This picture illustrates both the attractive and the discouraging aspects of this kind of evidence. It seems highly suggestive, but its history is difficult, if not impossible, to trace out. We would suggest, however, that the presence of Vedic cosmological themes in many widely separated cultures throughout the world does provide evidence for the existence of a single culture in the remote past that widely disseminated these themes.

3.b.4. The Principle of Correspondence

Thus far we have developed the idea that the earth of our experience is a small globe and simultaneously a part of a region called Bhārata-varṣa in a larger, higher-dimensional structure called Bhū-maṇḍala. We have proposed that the connection between the earth globe and Bhū-maṇḍala is higher-dimensional. Since this idea is very foreign to the Western way of thinking, we will devote this subsection to a discussion of further examples from the *Bhāgavatam* indicating that this earth (and India in particular) is linked with a higher level of reality. To borrow a phrase from modern physics, we can speak of this idea of a higher-dimensional connection as the principle of correspondence linking our familiar earth globe with the domain described in the Vedic literature.

There are many references in the *Bhāgavatam* indicating that in previous ages many activities of demigods and great *ṛṣis* were regularly carried out on this earth. These include the following:

(1) Trita Muni, who became one of the seven sages in the Varuṇaloka, came from the western countries of this earth (SB 1.9.6–7p).

(2) Inhabitants of the Vāyuloka (airy planets) were invited to expedite the cooking work at the sacrifice of Mahārāja Marutta. (Also, a golden mountain peak belonging Mahārāja Marutta is located somewhere in the Himalayas.) (SB 1.12.33p)

(3) Viśvāvasu, the leader of the Gandharvas, fell from his *vimāna* (airplane) upon seeing Devahūti playing ball on her palace roof. This took place in India (SB 3.22.17).

(4) Atri Muni performed austerities in a valley of Ṛkṣa Mountain near the river Nirvindhyā in India (SB 4.1.17).

(5) The sacrifice in which Dakṣa offended Lord Śiva took place at the confluence of the Ganges and the Yamunā (SB 4.2.35).

(6) Parvatī, the wife of Lord Śiva, took birth as the daughter of the Himalayas (SB 4.7.58–59).

(7) Svāyambhuva Manu ruled from Brahmāvarta, which is located in India where the river Sarasvatī flows toward the east (SB 4.19.1).

(8) Indra became intoxicated on *soma-rasa* at Mahārāja Gaya's sacrifice (SB 5.15.12).

These items all indicate that in the past this earth was the setting for many activities that lie beyond the range of our present senses. In the *Bhāgavatam* (including both the Sanskrit texts and Śrīla Prabhupāda's purports) these activities are described from the viewpoint of persons whose sensory level is higher than that of ordinary people of today, and thus they are presented as normal, day-to-day affairs. In the *Caitanya-caritāmṛta* there is evidence indicating that similar activities are still taking place on the earth today. For example, CC ML 9.174–77 describes a meeting that took place between Lord Caitanya and Lord Śiva on the hill of Śrī Śaila in south India. It is pointed out that Lord Śiva and Devī lived on that hill, along with Lord Brahmā and all the demigods. In this description, however, it is clear that this was not visible to the general human population.

In KB, p. 494 we read that the dowry of King Nagnajit's daughter included 90,000,000 horses and "a hundred times more slaves than horses." Modern scholars use statements like this as an excuse to reject Vedic scripture as "Hindu mythology," or utterly irresponsible fantasy. However, as we have already suggested, their interpretation is contradicted by the abundant evidence indicating the Vedic literature's gravity and seriousness of purpose. We suggest that these very large numbers refer to activities taking place on a higher earthly domain, which was experienced by the people of those times (the late Dvāpara-yuga).

In many cultures around the world we find the idea that in an earlier age people had direct contact with higher realms and their inhabitants (SH). This direct contact is often thought to have been broken in the distant past by a fall,

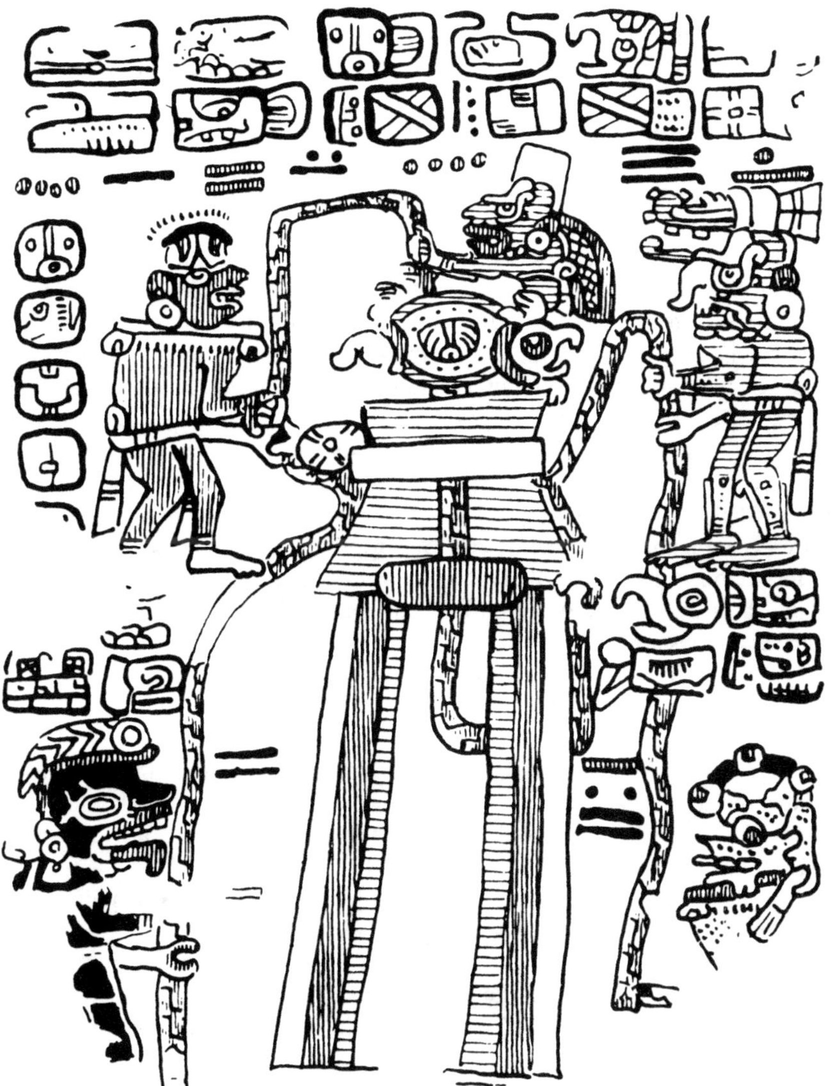

Figure 14. An illustration from the *Maya Codex Tro-Cortesianus* that may represent the churning of the milk ocean. The picture includes a tortoise, a central churn, and a serpent rope being pulled by what may be demigods and *asuras*.

which consigned human beings to a life of struggle in a state of cosmic alienation. The fall of Adam and Eve in Judeo-Christian tradition is an example of this. The Vedic literature, however, can be thought of as being written from a pre-fall perspective. Although this literature describes the degradation of human society in the Age of Kali, it generally describes activities and events taking place in societies where communication with higher-dimensional realms was taken for granted.

In SB 6.10.16p Śrīla Prabhupāda comments that the battle between Indra and Vṛtrāsura took place not by the Narmadā River in India, as one might surmise from the text, but by its celestial counterpart. He points out that "the five sacred rivers in India—the Gaṅgā, Yamunā, Narmadā, Kāverī, and Kṛṣṇā—are all celestial. Like the Ganges River, the Narmadā River also flows in the higher planetary systems." For this to be possible, the connection between the celestial river and the earthly river that we can directly see must be higher-dimensional.

Likewise, in SB 3.21.25p Śrīla Prabhupāda points out that Brahmāvarta, where Svāyambhuva Manu ruled, is said by some to be a place in India and by others to be a place in Brahmaloka. He says, "There are many places on the surface of this earth which are also known in the higher planetary systems; we have places on this planet like Vṛndāvana, Dvārakā, and Mathurā, but they are also eternally situated in Kṛṣṇaloka." Thus, a place in India on this earth may correspond on a higher-dimensional level to part of Brahmaloka.

In a number of places, Śrīla Prabhupāda cites traditions identifying features of the earth with features of Bhū-maṇḍala and the higher planets in general. Some examples are:

(1) "*Bhauma-svarga* [which corresponds to the eight *varṣas* of Jambūdvīpa other than Bhārata-varṣa] is sometimes accepted as the tract of land in Bhārata-varṣa known as Kashmir" (SB 5.17.11p).

(2) It is said that Śivaloka is "supposed to be situated near the Himalaya Mountains" (SB 4.24.22p).

(3) The Yakṣas (who are associated with the demigod Kuvera) are identified as Himalayan hill tribes like the Tibetans (SB 4.10.5p).

(4) The words *ā-mānasa-acalāt*, meaning "up to Mānasa Mountain," are translated as referring to the Arctic region (SB 4.16.14).

(5) "*Sapta-dvīpa* refers to the seven great islands or continents on the surface of the globe: (1) Asia, (2) Europe, (3) Africa, (4) North America, (5) South America, (6) Australia, and (7) Oceania" (SB 4.21.12p). Similar statements are made in SB 3.21.2p and TLC, p. 80.

We suggest that identifications of this kind either refer directly to higher-dimensional associations between earthly and celestial locations, or else they refer to traditions that have arisen because of ancient experience of the earth as a higher realm. Thus, Lord Śiva is always associated with the Himalayas, and in the Vedic literature there are many stories about him that take place in a Himalayan setting. It is therefore natural to think of the Himalayas as the place of Lord Śiva, and he may indeed be especially accessible there to advanced yogis. Of course, we cannot simply regard Śivaloka or Sapta-dvīpa as places in the three-dimensional earthly realm of our ordinary experience.

The astronomical *siddhāntas* also contain passages identifying features of Bhū-maṇḍala with parts of the earth globe. Thus the *Sūrya-siddhānta* describes Mount Meru as a small mountain at the North Pole, and the *Siddhānta-śiromaṇi* places the seven *dvīpas* in the Southern Hemisphere. In his purports to CC AL 5.111 and CC ML 20.218, Śrīla Prabhupāda cites the *Siddhānta-śiromaṇi*'s description of the seven *dvīpas*. Since Śrīla Bhaktisiddhānta Sarasvatī Ṭhākura also cites this description in his *Anubhāṣya* commentary on these verses of *Caitanya-caritāmṛta*, we will reproduce it here:

> Most learned astronomers have stated that Jambūdvīpa embraces the whole northern hemisphere lying to the north of the salt sea; and that the other six *dvīpas* and the seven seas . . . are all situated in the southern hemisphere.
>
> To the south of the equator lies the salt sea, and to the south of it the sea of milk, . . . where the omnipresent Vāsudeva, to whose lotus feet Brahmā and all the gods bow in reverence, holds his favorite residence.
>
> Beyond the sea of milk lie in succession the seas of curds, clarified butter, sugar cane juice, and wine; and, last of all, that of sweet water, which surrounds Vadavānala. The Pātāla *lokas*, or infernal regions, form the concave strata of the earth [SSB1, p. 116].

We should note that these verses of *Siddhānta-śiromaṇi* describe a correspondence between the earth globe and Bhū-maṇḍala that can be expressed in mathematical form. The points on the plane of Bhū-maṇḍala can be mapped onto the earth globe by a stereographic projection. This is a standard kind of map projection, in which countries on the curved surface of the earth are represented on a flat plane.

In this particular case, one can use a modified polar stereographic projection, which sends the North Pole of the earth to the center point on the plane and sends circles of latitude on the earth to ever-widening concentric circles on the plane. It is possible to set up such a projection so that

(1) The path of the sun in Puṣkaradvīpa maps to the tropic of Capricorn (see Section 3.d).

(2) The six *dvīpas* surrounding Jambūdvīpa map to bands along parallels of latitude in the Southern Hemisphere.

(3) The equator cuts the salt ocean between Jambūdvīpa and Plakṣadvīpa in half. Thus Jambūdvīpa lies in the Northern Hemisphere.

(4) The base of Mount Meru maps to the Arctic Circle. Thus Mount Meru

corresponds to the "land of the midnight sun," north of the Arctic Circle.

This correspondence agrees with the description of the *dvīpas* in the *Siddhānta-śiromaṇi*, and it agrees with the account given in the *Sūrya-siddhānta* of the life of the demigods on Mount Meru. There it is stated that the demigods experience days and nights of six months each, and that their dawn and evening occur at the times of the vernal and autumnal equinoxes (SS, p. 81). This, of course, is the situation at the North Pole.

The question is, What is the meaning of this mapping between Bhū-maṇḍala and the earth globe? It is not possible for us to take it as a literal description of the earth, since the continents in the Southern Hemisphere are not at all arranged in concentric rings. It may be that this mapping refers to actual higher-dimensional connections between parts of this earth and parts of Bhū-maṇḍala. This is suggested by the fact that Śrīla Bhaktisiddhānta Sarasvatī refers to it, and Śrīla Prabhupāda, following in disciplic succession, does also.

However, since the authors of the astronomical *siddhāntas* often expressed doubts about Purāṇic cosmology, it seems likely that for them, at least, the mapping was simply an artificial attempt to force this cosmology into a three-dimensional framework and thereby make sense out of it. We therefore suggest that although historical Indian astronomers such as Bhāskarācārya were carrying on a genuine Vedic tradition of astronomy, their understanding of Vedic cosmology was nonetheless imperfect. They did not understand the higher-dimensional nature of structures such as Bhū-maṇḍala, and they consequently focused their attention on those features of Vedic astronomy that can be readily understood in three-dimensional terms.

In recent centuries, many Vaiṣṇavas have also experienced perplexity in their efforts to understand the relationship between Bhū-maṇḍala and the earth globe of our direct experience. This is shown in Appendix 1, where we reproduce a discussion of this relationship by the Vaiṣṇava commentator Vaṁśīdhara. If the existing Vedic literature consists of materials dating to an era in which people had direct experience of higher-dimensional reality, then it is not surprising that many statements in it are bewildering from our gross sensory perspective. It is therefore reasonable to follow the example of the *ācāryas* and simply receive these statements with faith. If this is done, then further insight may come in due course of time. (In contrast, the approach of skeptical rejection is not likely to lead to further study and insight.)

We will end this subsection by noting another correspondence principle involving Vedic cosmology—the principle of correspondence between microcosm (the body) and macrocosm (the universe and the universal form). In SB 5.23cs there is the statement that "*yogīs* worship the Śiśumāra planetary system, which is technically known as the *kuṇḍalini-cakra*." It appears that *yogīs* in meditation would identify the central axis of the universe (which we will discuss in Chapter 4) with the series of *cakras* in the spinal column. By moving their life airs up the series of *cakras*, they would prepare their subtle bodies to travel up the axis of the universe to Brahmaloka. This basic idea appears in mystical traditions throughout the world, but it would take us too far afield to discuss it further here (again, see SH).

3.C. PLANETS AS GLOBES IN SPACE

In the pastime of Lord Varāha's lifting the earth from the ocean, the earth is frequently depicted by artists as our familiar earth globe. However, the Sanskrit verses of *Śrīmad-Bhāgavatam* describing this pastime do not use any words denoting a sphere when referring to the earth, and the *Viṣṇu Purāṇa* indicates that Lord Varāha lifted Bhū-maṇḍala as a whole. The relevant passage states that after lifting the earth from the waters, Lord Varāha divided it into seven great continents, as it was before, thus indicating that the earth that was lifted included the seven *dvīpas* of Bhū-maṇḍala (VP, p. 65). The Vaiṣṇava commentator Vaṁśīdhara, in his commentary on SB 5.20.38, also points out that the earth lifted by Lord Varāha is Bhū-maṇḍala (see Appendix 1).

In the Fifth Canto the earth is directly described as the vast disc of Bhū-maṇḍala. The word *bhū-golam*, or "earth-globe," generally refers to the sphere of the universe, and the *Bhāgavatam* seems to make no direct reference to the earth as a small globe. However, the astronomical *siddhāntas* do explicitly describe the earth as a small globe, and the following verse in the Fifth Canto can be interpreted as referring to the earth as a sphere:

> People living in countries at points diametrically opposite to where the sun is first seen rising will see the sun setting, and if a straight line were drawn from a point where the sun is at midday, the people in countries at the opposite end of the line would be experiencing midnight. Similarly, if people residing where the sun is setting were to go to countries diametrically opposite, they would not see the sun in the same condition [SB 5.21.8–9].

We have argued that the earth was understood to be a sphere in Vedic times, and that it was also understood to be part of Bhū-maṇḍala. It is therefore natural to ask whether or not the other parts of Bhū-maṇḍala also correspond to spheres in some sense. In fact, Śrīla Prabhupāda frequently refers to the idea of planets as globes floating in space. Since this point is quite important, we shall quote a number of his statements at length:

(1) "The earth floats in space among many millions of other planets, all of them bearing huge mountains and oceans. It floats because Kṛṣṇa enters into it, as stated in *Bhagavad-gītā* (*gām āviśya*), just as He enters the atom" (TQK, p. 122).

(2) "Seated on His chariot with Arjuna, Kṛṣṇa began to proceed north, crossing over many planetary systems. These are described in the *Śrīmad-Bhāgavatam* as Saptadvīpa. *Dvīpa* means 'island.' These planets are sometimes described in the Vedic literature as *dvīpas*. The planet on which we are living is called Jambūdvīpa. Outer space is taken as a great ocean of air, and within that great ocean of air there are many islands, which are the different planets. In each and every planet there are oceans also. In some of the planets the oceans are of salt water, and in some of them there are oceans of milk. In others there are oceans of liquor, and in others there are oceans of ghee or oil" (KB, pp. 855–56). Similar remarks are made in KB, p. 12.

(3) "The planets are called *dvīpas*. Outer space is like an ocean of air. Just as there are islands in the watery ocean, these planets in the ocean of space are called *dvīpas*, or islands in outer space" (CC ML 20.218p). This purport begins with a quotation of the Sanskrit verses from *Siddhānta-śiromaṇi* describing the seven *dvīpas* of Bhū-maṇḍala, and thus Śrīla Prabhupāda clearly does not limit the *dvīpas* to the Southern Hemisphere.

(4) "Sometimes the planets in outer space are called islands. We have experience of various types of islands in the ocean, and similarly the various planets, divided into fourteen *lokas*, are islands in the ocean of space. As Priyavrata drove his chariot behind the sun, he created seven different types of oceans and planetary systems, which altogether are known as Bhū-maṇḍala, or Bhūloka" (SB 5.1.31p).

(5) "According to Vedic understanding, the entire universe is regarded as an ocean of space. In that ocean there are innumerable planets, and each planet is called a *dvīpa*, or island" (SB 8.19.19p).

(6) "Only under certain conditions do the planets float as weightless balls in the air, and as soon as these conditions are disturbed, the planets may fall down into the Garbhodaka Ocean, which covers half the universe. The other half is the spherical dome within which the innumerable planetary systems exist. The floating of the planets in the weightless air is due to the inner constitution of the globes" (SB 2.7.1p).

(7) In SB 2.7.13p, 1.3.41p, and 3.15.2p it is indicated that the universe contains millions of planets, and that many are not visible to the naked eye.

In these passages Śrīla Prabhupāda refers to the seven *dvīpas* of Bhū-maṇḍala as a planetary system consisting of many globes floating in space. He compares outer space to an ocean of air and interprets the word *dvīpa* to mean an island hovering in that airy ocean. Since the *Bhāgavatam* does not specifically refer to the *dvīpas* as separated globes, this naturally gives rise to the question, Is the *Bhāgavatam* giving a metaphorical description of the universe, and if so, then how far can we go in giving indirect interpretations to its statements? We note that passage (4) refers to a verse in which it is said that Mahārāja Priyavrata created the seven *dvīpas* and oceans of Bhū-maṇḍala with the rims of his chariot wheels. We can easily see how a very large chariot could produce circular ruts that would become oceans and islands, but it is not so easy to see how it could produce systems of spherical planets.

In answer to the above question, we suggest that the statements of the *Bhāgavatam* can sometimes be given indirect interpretations, but this should be done very carefully in accordance with the overall meaning of the text and the tradition of *paramparā*. According to the Vedic literature, the universe is very difficult to understand, and a complete element-by-element description in the modern Western style is not possible. Any description can depict only a limited aspect of the total reality, and to do this the description must make use of familiar concepts and images. Thus to some extent any description of the universe must be indirect and metaphorical.

Whenever we read a statement and arrive at some understanding of it, we are necessarily interpreting it in the context of many underlying assumptions, some of which we may hold unconsciously. Thus, as we have already pointed out, a literal reading of a text is also an interpretation, and it may be an incorrect one. What then is the right way to understand a text? We suggest that this can be properly done only if one makes a sincere effort to enter into the spirit of the text as a whole and tries to realize the meaning intended by its author. Since the author is invariably writing in the context of some tradition, this also means immersing oneself in that tradition in an effort to assimilate its world view.

Thus far we have been presenting a picture of Vedic cosmology based on the observation that the Vedic literature is using familiar three-dimensional imagery to describe an inherently non–three-dimensional material and spiritual reality. According to this interpretation, the simple image of the disc of Bhū-maṇḍala has been used to describe a higher-dimensional situation in which the earth can be seen in a variety of ways at different levels of sensory perception. The simple image of travel in outer space has likewise been used to describe modes of yogic travel that defy understanding in three-dimensional terms.

If we proceed with this interpretation of the Vedic world view, then one way to understand the idea of the *dvīpas* as islands in space is as follows: As the earth, which is part of Bhū-maṇḍala, appears to be a small globe to our ordinary senses, so various parts of Bhū-maṇḍala (and other regions of the universe) may also be experienced as globes floating in space by beings with certain levels of sensory development. On the basis of logic alone, we would offer this idea only as a tentative conjecture. However, since Śrīla Prabhupāda is writing in accordance with the *paramparā* tradition, we suggest that this idea of Bhū-maṇḍala as a system of floating planetary globes must be in accord with the Vedic literature as a whole. It simply represents the appearance of Bhū-maṇḍala at one sensory level.

3.D. THE ORBIT OF THE SUN

"The universe is like a tree with the roots being upwards. The polestar which is situated within the Asking question star constellation is the root. The universe is pivoting around the pole star. That is one movement. The second movement is that the sun is revolving around the universe, as if it were going around the tree" (letter from Śrīla Prabhupāda to Svarūpa Dāmodara dāsa, November 21, 1975).

In this section we will discuss what the *Bhāgavatam* has to say about the movement of the sun, and then we will use this information to develop the two hypotheses about the projection of Bhū-maṇḍala on the sky that we mentioned in Section 3.b.2. As Śrīla Prabhupāda indicates in the above quote, the sun moves with respect to the reference frame of this earth in two different ways. The most noticeable motion is the daily rotation of the sun from east to west around the earth, which produces the phenomena of day and night. The stars and planets also participate in this motion, and they all appear to revolve once

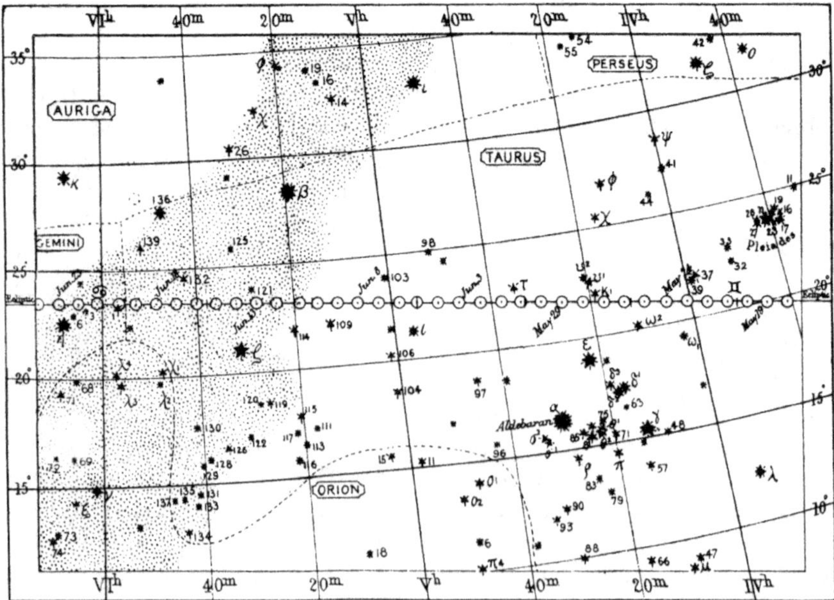

Figure 15. The part of the ecliptic traversed by the sun from about May 20 to June 21.

per day around a fixed axis passing through the polestar.

The second motion is the slow movement of the sun from west to east with respect to the stars. This movement takes place along the celestial great circle known as the ecliptic. To visualize this, consider that stars are present during the day, but we cannot see them due to the brilliant sunlight. If we could see them, we would see that on a particular day the sun is surrounded by certain stars. A day later, the sun will have shifted eastward relative to these stars by about one degree. Day by day the sun continues to shift until it completes one revolution around the ecliptic in one year. In the course of this revolution it passes by the various star constellations of the zodiac, which are laid out along the ecliptic (see Figs. 10 and 15).

The ecliptic is tilted at a 23.5° angle to the celestial equator, which is perpendicular to the polar axis. Thus, as the sun moves along the ecliptic, it moves toward the celestial north pole (the polestar) for half the year, and it moves toward the celestial south pole for the other half. When it is north of the celestial equator, days are longer than nights in the Northern Hemisphere, and the opposite is true in the Southern Hemisphere. This situation is reversed when the sun is south of the celestial equator.

3.d.1. The Ecliptic as the Projection of Bhū-maṇḍala on the Celestial Sphere

Our first hypothesis is that the projection of Bhū-maṇḍala on the celestial sphere coincides closely with the ecliptic. The basic argument for this goes as follows: In the Fifth Canto we read that the sun orbits Mount Meru, moving above a ring-shaped mountain in Bhū-maṇḍala called Mānasottara. This ring is centered on Mount Meru, and it has a circumference of 95,100,000 *yojanas* (SB 5.21.7). The radius of this ring is about 15,750,000 *yojanas*, and the height of the sun above Bhū-maṇḍala is 100,000 *yojanas* (SB 5.23.9p). (Here the *Bhāgavatam* is using 3 as an approximation for π.) This means that the distance from the sun to an observer on the earth is much greater (by a factor of 157.5) than the distance from the sun to the plane of Bhū-maṇḍala.

Therefore, the part of Bhū-maṇḍala that lies directly underneath the sun at any given time must seem to be very close to the sun from the point of view of an observer on the earth. In other words, that part of Bhū-maṇḍala must project to a point on the celestial sphere that is very close to the location of the sun. We know where the sun is on the celestial sphere at any given time. So, if we can find out where the sun is in Bhū-maṇḍala at successive moments in time, then we can see where Bhū-maṇḍala falls on the celestial sphere.

The following statements from the *Bhāgavatam* indicate that the sun makes one circuit around Mānasottara Mountain per year, and that the sun is due north of Mount Meru when it moves farthest to the north on the celestial sphere. (This is called the summer solstice, and it occurs in June.)

> [1] Encircling Sumeru Hill on his chariot, the sun-god illuminates all the surrounding planetary systems. However, when the sun is on the northern side of the hill, the south receives less light, and when the sun is in the south, the north receives less [SB 5.1.30]. [Śrīla Prabhupāda comments,] According to *Jyotir Veda*, the science of astronomy in the Vedic literature, the sun moves for six months on the northern side of Sumeru Hill and for six months on the southern side. We have practical experience on this planet that when there is summer in the north there is winter in the south and vice versa.
>
> [2] In the chariot of the sun-god, the sun travels on the top of the [Mānasottara] mountain in an orbit called Saṁvatsara, encircling Mount Meru. The sun's path on the northern side is called Uttarāyaṇa, and its path on the southern side is called Dakṣiṇāyana. One side represents a day for the demigods, and the other represents their night [SB 5.20.30].
>
> [3] Śukadeva Gosvāmī continued: My dear King, as stated before, the learned say that the sun travels over all sides of Mānasottara Mountain in a circle whose length is 95,100,000 *yojanas* [760,800,000 miles]. On Mānasottara Mountain, due east of Mount Sumeru, is a place known as Devadhānī, possessed by King Indra. Similarly, in the south is a place known as Saṁyamanī, possessed by Yamarāja, in the west is a place known as Nimlocanī, possessed by Varuṇa, and in the north is a place named Vibhāvarī, possessed by the moon-god. Sunrise, midday, sunset, and midnight occur in all those places according to specific times, thus engaging all living entities in their various occupational duties and also making them cease such duties [SB 5.21.7].

Passages (1) and (2) indicate that the sun takes one year to make a complete circuit around Mānasottara Mountain. From passage (3) we see that on the plane of Bhū-maṇḍala, the directions north, south, east, and west are laid out in the same way as on a flat Mercator projection of the earth's surface. In Jambūdvīpa, Bhārata-varṣa is to the south of Mount Meru, and Uttarakuru-varṣa is to the north. Saṁyamanī is much further to the south, on the ring-shaped *dvīpa* of Puṣkaradvīpa, and Vibhāvarī is located on this *dvīpa* equally far to the north of Mount Meru. It would seem that the sun spends half the year in the part of its orbit lying to the north of Mount Meru, and half the year in the part lying to the south. This yearly circuit of the sun through Bhū-maṇḍala provides a simple explanation for the many statements in the *Bhāgavatam* indicating that the demigods' day (24 hours) lasts for one earthly year.

The motion of the sun through Bhū-maṇḍala is described as follows in the *Bhāgavatam*:

> The chariot of the sun-god has only one wheel, which is known as Saṁvatsara. The twelve months are calculated to be its twelve spokes, the six seasons are the

> sections of its rim, and the three *cātur-māsya* periods are its three-sectioned hub. One side of the axle carrying the wheel rests upon the summit of Mount Sumeru, and the other rests upon Mānasottara Mountain. Affixed to the outer end of the axle, the wheel continuously rotates on Mānasottara Mountain like the wheel of an oil-pressing machine [SB 5.21.13].

According to this description, we can imagine the sun moving in a circle around Bhū-maṇḍala in much the same way as a horse-drawn chariot moves around a race track. In discussing this verse, we should comment on the use of metaphor in the *Bhāgavatam*. One example of metaphorical description is the story of the city of nine gates entered by King Purañjana. There the different gates of the city symbolize different bodily senses. In the verse we have just quoted, the different parts of the wheel of the sun-god's chariot similarly symbolize different divisions of the year. Thus one might take this verse as a metaphorical description of the movement of the sun during the year. As a general rule, since the purpose of metaphor is to increase understanding and not to obscure it, such indirect interpretation is justified only if the intended metaphorical meaning is transparently clear. One should not devise a metaphorical interpretation simply to replace a clear direct meaning.

Whether the Saṁvatsara wheel should be taken metaphorically or not, the verse clearly states that the sun is moving only a short distance above Bhū-maṇḍala. The comparison with an oil-pressing machine indicates that the chariot of the sun is always directly in contact with the upper surface of the ring-shaped Mānasottara Mountain. The identification of the wheel with the year is also consistent with the view that the sun takes one year to make a complete circuit of Bhū-maṇḍala.

When the path of the sun is projected onto the sky from our vantage point, it lies in the zodiac. According to SB 5.22.5, "Passing through twelve months on the wheel of time, the sun comes in touch with twelve different signs of the zodiac and assumes twelve different names according to those signs. The aggregate of those twelve months is called a *saṁvatsara*, or an entire year." If the circuit of the sun through Bhū-maṇḍala (called "Saṁvatsara" in (2) above) takes one year, then the successive parts of Bhū-maṇḍala visited by the sun must correspond to the successive parts of the zodiac lying along the ecliptic. From this we conclude that when Bhū-maṇḍala is projected in the sky, it must lie on the ecliptic, with the northernmost part of Mānasottara Mountain corresponding to the summer solstice.

This means that Bhū-maṇḍala remains stationary with respect to the stars, with the signs Gemini and Cancer (Mithuna and Karkaṭa) in the direction of Vibhāvarī to the north of Mount Meru, and the signs Capricorn and Sagittarius (Makara and Dhanur) in the direction of Saṁyamanī to the south of Mount Meru (see SB 5.21cs). Since the stars rotate once per day around the polar axis, it must be that Bhū-maṇḍala also rotates once per day around this axis. This in turn implies that there is a relative rotation between Bhū-maṇḍala and the earth of our experience. It is not correct to assume naively that this earth and the rest of Bhū-maṇḍala form a single rigid plate.

Now, this conclusion might be regarded as a drawback to the hypothesis that Bhū-maṇḍala corresponds to the ecliptic. It could be argued that the "earth," or Bhū, is motionless according to the Vedic literature. If Bhū-maṇḍala rotates daily with the stars and planets, then its system of directions—north, south, east, and west—also rotates and therefore does not correspond to our earthly system of directions. It could also be argued that in the statement that the sun spends half the year to the north of Mount Meru, "north" should be interpreted as meaning the north of the celestial sphere, and "Mount Meru" should be taken as the equator of this sphere.

In response to these arguments, one can reply that if Bhū-maṇḍala is indeed a system of spherical planets floating in space, then why shouldn't it rotate daily around the celestial pole along with the other stars and planets? We can see how the yearly circling of the sun through this system would produce a day of one year for the higher beings on each of these planets, if they do not rotate about their own axes. In any event, whether Bhū-maṇḍala rotates or not, its system of directions cannot correspond to the earthly system: The earthly north, south, east, and west point in different directions at different points on the spherical earth, while a set of directions on a plane have the same orientation at every point. (For example, at the North Pole every direction is south, but at Mount Meru the four directions are clearly defined.)

3.d.2. The Celestial Equator as the Projection of Bhu-maṇḍala on the Celestial Sphere

These objections to our first hypothesis suggest a second hypothesis about the projection of Bhū-maṇḍala. This is that the projection of Bhū-maṇḍala on the sky coincides with the celestial equator. This implies that the plane of Bhū-maṇḍala is parallel to the earth's surface at the poles. At the North Pole, the sun is visible in the sky for half the year. It rises above the horizon at the time of the vernal equinox and spirals slowly up into the sky, making one turn per day. At the time of the summer solstice it reaches a high point of 23.5° above the horizon, and then slowly spirals down, reaching the horizon again at the autumnal equinox. According to this hypothesis, this is how the behavior of the sun would appear to a hypothetical observer standing on one of the *dvīpas* of Bhū-maṇḍala.

To back up this hypothesis, we first note the following verses, which seem to contradict the idea that the sun makes one circuit through Bhū-maṇḍala per year:

> When the sun travels from Devadhānī, the residence of Indra, to Saṁyamanī, the residence of Yamarāja, it travels 23,775,000 *yojanas* [190,200,000 miles] in fifteen *ghaṭikās* [six hours].
>
> From the residence of Yamarāja the sun travels to Nimlocanī, the residence of Varuṇa, from there to Vibhāvarī, the residence of the moon-god, and from there again to the residence of Indra. In a similar way, the moon, along with the other stars and planets, becomes visible in the celestial sphere and then sets and again becomes invisible.
>
> Thus the chariot of the sun-god, which is *trayīmaya*, or worshiped by the words *oṁ bhūr bhuvaḥ svaḥ*, travels through the four residences mentioned above at a speed of 3,400,800 *yojanas* [27,206,400 miles] in a *muhūrta* [SB 5.21.10–12].

Here we should note some technical details. First, 15 *ghaṭikās* equals one fourth of a day, and 23,775,000 *yojanas* is indeed one fourth of the 95,100,000-*yojana* circumference of Mount Mānasottara. The figure of 3,400,800 *yojanas* per *muhūrta* is more difficult to interpret. Normally, there are 30 *muhūrtas* in a day. However, SB 3.11.8 implies that standards of 24 or 28 *muhūrtas* per day were also used. If we use 28, we see that 28 times 3,400,000 is 95,200,000. Also, in SB 5.21.19 the sun is said to move 2,000.5 *yojanas* per moment, or *kṣaṇa*. This is consistent with 3,400,800 *yojanas* per *muhūrta* if we use 1,700 moments per *muhūrta*. (SB 3.11.7–8 indicates 2,250 *kṣaṇas* per *muhūrta*.)

All of these verses say that the sun makes one circuit through Bhū-maṇḍala in a day. If we take this to be the case, then on each day there will be a time when the sun is located above Vibhāvarī, the residence of the moon-god on Mount Mānasottara. At this time on successive days, the sun will occupy a succession of different positions along the ecliptic. The ecliptic itself makes one rotation per sidereal day around the polar axis, and in one solar day it makes slightly more than one rotation. (A sidereal day is measured from star-rise to star-rise, and a solar day is measured from sunrise to sunrise.) If we argue, as before, that Vibhāvarī must be close to the sun on the celestial sphere when the sun passes over it, then it follows that the projection of Vibhāvarī on the celestial sphere must make one orbit per year through the ecliptic.

Combining this motion with the motion of the ecliptic on successive days, and assuming that the sun rotates around Mānasottara Mountain once per solar day, we find that the position of Vibhāvarī on successive days moves slowly up and down between the uppermost and lowermost limits of the ecliptic. By applying this reasoning to a number of other locations in Bhū-maṇḍala, we arrive at the following picture: Bhū-maṇḍala itself moves up and down parallel to the celestial equator in a cyclic motion taking one year to complete.

This is a very strange motion, and it contradicts the assumption that the earth is located in the plane of Bhū-maṇḍala. Clearly something has to give here. One possibility is to relax the requirement that the sun is always close to Bhū-maṇḍala (relative to its distance from us). This allows us to place Bhū-maṇḍala in the plane of the celestial equator. We now suppose that the sun moves up and down with respect to Bhū-maṇḍala in a yearly cycle while also circling Bhū-maṇḍala once per day. This gives the pattern of solar motion that is seen at the North Pole.

This is our second hypothesis. Although it conforms with SB 5.21.10–12, it does have the drawback that it allows the sun to move quite far from the plane of Bhū-maṇḍala. According to the *Bhāgavatam*, the distance from Jambūdvīpa to Mānasottara Mountain is 126 million miles. Thus at the summer solstice, when the sun is 23.5° above the celestial equator, our second hypothesis implies that the sun is about 54,786,000 miles above Bhū-maṇḍala. At the vernal and autumnal equinoxes it is in the plane of Bhū-maṇḍala, and at the winter solstice it is 54,786,000 miles *below* this plane. This does not agree very well with the descriptions of the sun's motion around Mount Meru on a chariot comparable to an oil-pressing machine. It also does not agree with the story of Mahārāja Priyavrata, who followed the sun in a chariot that moved over the plane of Bhū-maṇḍala and created the seven oceans by making ruts with its wheels.

The point can also be made that the daily clockwise (or east-to-west) motion of the sun is due to the *dakṣiṇāvarta* wind, according to SB 5.21.8–9. In general, the movement of the planets around the polar axis is attributed to a wind (SB 5.23.3). If the daily motion of the sun is also due to this wind, then one can suggest that the sun's yearly counter-clockwise motion could be due to the movement of the sun's chariot through Bhū-maṇḍala. This interpretation supports our first hypothesis, and it is confirmed by the following remark by Śrīdhara Svāmī in his commentary on SB 5.21.8–9:

> Although leftward movement, facing the constellations, is their own motion [*svagatya*], the luminaries [sun, moon, etc.] move around Meru to the right daily, being blown by the *pravaha* wind, due to the power of the [*kāla*] *cakra*.

Here the *svagatya*, or "own motion," of the sun must be its yearly motion around the ecliptic, since this movement is to the left (if one faces the constellations of the zodiac) and the daily motion due to the wind is to the right. Thus the sun's chariot should be moving counter-clockwise around Mount Meru. (This assumes that the observer is, say, in northern India, where the constellations of the zodiac are to the south. In the southern hemisphere, south of the tropic of Capricorn, everything would be reversed, but the same conclusion about the movement of the sun would hold.)

We suggest that further research will be necessary for us to give a final conclusion regarding the celestial orientation of Bhū-maṇḍala in Vedic cosmology.

Here we tentatively propose that the Fifth Canto of the *Bhāgavatam* is presenting a combined description of the two types of solar motion. Bhū-maṇḍala is being used as the underlying framework in each description, and thus a contradictory picture of its position seems to emerge. We note that a combined description of the two forms of solar motion is explicitly made in SB 5.21.8–9 and SB 5.22.1–2, and the idea of relative motion is introduced. These verses speak of the sun-god circling Mount Meru with the mountain on his left *and* on his right. Unfortunately, however, they do not specify which motion is actually taking place, relative to the plane of Bhū-maṇḍala.

In spite of these ambiguities, it does appear that the intent of the *Bhāgavatam* is to present Bhū-maṇḍala as an actually-existing, disc-shaped domain. We have suggested that its location in space must be related to the geocentric orbit of the sun. In Section 4.b we will also argue that its location can be related to the orbits of the moon and other planets. This argument will provide further evidence in support of the hypothesis connecting Bhū-maṇḍala with the ecliptic.

We would finally like to draw attention to the statement in SB 5.21.11 that "in a similar way, the moon, along with the other stars and planets, becomes visible in the celestial sphere and then sets and again becomes invisible." This statement seems to be another indirect reference to the spherical shape of the earth planet: Since the luminaries are rotating once per day around this sphere, they seem to rise and set daily at any given place (between the Arctic and Antarctic circles).

The Vertical Dimension

Thus far we have discussed the plane of Bhū-maṇḍala, and we have largely confined our attention to the two-dimensional region of space that this plane defines. In addition to this plane, which we can think of as horizontal, Vedic cosmology also has a vertical dimension. We naturally tend to define the direction "up" as meaning "away from the earth's center," and when we speak of the distance of an object from the earth, we mean its distance from this center. In Vedic cosmology, however, "up" means "toward celestial north, in a direction perpendicular to the plane of Bhū-maṇḍala," and "down" means the opposite direction. The distance of an object from the earth in Vedic cosmology is the length of a perpendicular line from the object to this plane. As we shall see, this concept of distance is important for our understanding of the relative distances of the sun and the moon in Vedic cosmology.

4.A. THE TERMINOLOGY OF THREE AND FOURTEEN WORLDS

Along this vertical direction, the universe is divided into three and also fourteen subdivisions. The three subdivisions are called the three worlds: lower, middle, and upper. These worlds are often referred to by the names Bhūḥ, Bhuvaḥ, and Svaḥ, as well as the names Pātāla, Martya, and Svarga (SB 3.11.28p). However, these two sets of names are not synonymous. Svaḥ and Svarga both denote the realm of the demigods, which lies above Bhū-maṇḍala. Bhūḥ or Bhūrloka refers to the earthly planetary system, including Bhū-maṇḍala and this earth (SB 4.20.35p), and Bhuvaḥ or Bhuvarloka refers to a planetary system lying between Bhūḥ and Svar (SB 2.5.40–41p). Apparently, human beings live in both the Bhūḥ and Bhuvaḥ systems (SB 1.9.45p).

Going from lowest to highest, the fourteen subdivisions are Pātāla, Rasātala, Mahātala, Talātala, Sutala, Vitala, Atala, Bhūrloka, Bhuvarloka, Svargaloka, Maharloka, Janaloka, Tapoloka, and Satyaloka. The word *Pātāla* is sometimes used to refer collectively to the seven lower planetary systems from Pātāla up to Atala. These are all described as discs lying below Bhū-maṇḍala and parallel to it. The words *Martya* and *Martyaloka* also designate the Bhūrloka system and refer to the fact that this system is a place of suffering and death. The six planetary systems from Bhuvarloka to Satyaloka are known as the higher planets. Śrīla Prabhupāda also uses the terminology "upward" planetary systems for Bhūrloka through Satyaloka, and "downward" planetary systems for Atala through Pātāla (SB 2.1.26p).

We have already noted that the three worlds—Pātāla, Martya, and Svarga—are also sometimes known as three kinds of Svargas, or heavenly regions (SB 5.17.11p). These three Svargas are explicitly defined as follows in the *Śrī Bṛhad-bhāgavatāmṛtam* of Śrīla Sanātana Gosvāmī: "(1) Vila-svarga: Atal, Bital, Sutal, Talātal, Mahātal, Rasātal, and Pātal. . . . (2) Bhauma-svarga: Jambu, Plaksha, Shalmali, Kusha, Crouncha, Shaka, and Puskara. . . . (3) Divya-svarga: the world of the devatās" (BB, p. 107). Here the three subdivisions Bila-svarga, Bhauma-svarga, and Divya-svarga correspond exactly to Pātālaloka, Martyaloka, and Svargaloka.

4.B. THE SEVEN PLANETS

There are seven traditional planets in the sky that are readily visible to human beings. These are the sun, the moon, Venus, Mercury, Mars, Jupiter, and Saturn. Of these, Śrīla Prabhupāda has specifically said that the moon belongs to Svargaloka, or "the third status of the upper planetary system," and the same is presumably true of the others (SB 2.5.40–41p). The moon and the sun are given a distinctive position among the planets of Svargaloka in SB 3.11.29–30, where it is said that after the three worlds are annihilated at the end of Brahmā's day, the sun and moon continue to exist. Śrīla Prabhupāda has pointed out that although the different planetary systems are described as lying in successive layers, like phonograph records in a stack, actually the planets of different types are mixed together:

> Regarding your question of the planetary systems, the planets are arranged in each universe in layers like the petals of a lotus. But in each layer there is mixed both heavenly, hellish, and middle planets. On the outside layer there are these three kinds of planets, on the middle layer there are the three kinds of planets, and on the innermost layer there are found these three kinds of planets. Above these layers, in the center, is the Brahmaloka, where Lord Brahmā, the creator, is residing. So the earth planet and the moon planet are both in the same layer, but the earth is a middle planet and the moon is a heavenly planet [letter to Rūpānuga dāsa, December 20, 1968].

This letter indicates that the moon is a heavenly planet, but suggests that it can occupy the same level in the vertical direction as the earth.

In the *Bhāgavatam* there are many stories that take place in Svargaloka, but these are rarely (if ever) set specifically on one of the seven planets. However, these planets played an important role in Vedic society because their visible motions were understood to be indicators of the course of events on the earth, both on the level of individuals and on the level of society as a whole. This, of course, is the subject matter of astrology, and we have already pointed out (in Chapter 1) that since astrology was regarded as very important in Vedic society, astronomy, and specifically the study of the motions of the seven planets, was also regarded as very important.

Although the *Bhāgavatam* gives a fairly detailed account of the movements of the sun, it gives only a relatively brief description of the movements of the other planets. The only information given about the positions of the planets is a list of their heights above Bhū-maṇḍala. Their horizontal positions over the plane of Bhū-maṇḍala are not mentioned. This list is given in Table 8.

The two most striking features of this list of planetary distances are (1) that the moon is listed as being higher than the sun, and (2) that the distances for the planets other than the moon are all much smaller than the values given to them by modern astronomers (see Table 1). To many people, this would seem to indicate that the *Bhāgavatam* is giving an extremely unrealistic account of the positions of the planets. However, this is not necessarily so.

The key point to consider here is that these distances are all *heights* of the planets above the plane of Bhū-maṇḍala. They are not distances along the line of sight from the earth to the planets. Let us therefore suppose that the distances of the planets from this earth along the plane of Bhū-maṇḍala might be much larger than the figures in Table 8.

TABLE 8

The Heights of the Planets Above Bhū-maṇḍala

Planet	Height above Bhū-maṇḍala
Sun	800,000
Moon	1,600,000
Venus	4,800,000
Mercury	6,400,000
Mars	8,000,000
Jupiter	9,600,000
Saturn	11,200,000

These figures, which are based on 8 miles per *yojana*, were obtained by using the planet-to-planet intervals from SB 5.22, plus the earth-to-sun distance given in SB 5.23.9p. The planetary heights listed in the verse translations in Chapter 22 are 800,000 miles higher than the figures in this table.

TABLE 9

The Maximum Distances the Planets Move from the Plane of the Ecliptic

Planet	Orbital Radius	Orbital Inclination	Maximum Distance from the Ecliptic
Sun	1.00 AU	0.000	0
Moon	238,000 miles	5.150	21,364
Venus	.72 AU	3.400	3,971,000
Mercury	.39 AU	7.167	4,525,000
Mars	1.52 AU	1.850	4,564,000
Jupiter	5.20 AU	1.317	11,115,000
Saturn	9.55 AU	2.480	38,431,000

Here modern Western data (EA) is used to compute the maximum distance in miles that each planet travels from the plane of the ecliptic in the course of its orbit. This is the average radius of the orbit times the sine of the inclination of the orbit to the ecliptic. Geocentric orbits were used for the sun and moon, and heliocentric orbits were used for the other planets. (1 AU = 93,000,000 miles.)

This is true in the case of the sun, since the distance from Jambūdvīpa to Mount Mānasottara is about 126,000,000 miles, using 8 miles per *yojana*. Using our smaller figure from *Sūrya-siddhānta* of 5 miles per *yojana*, this distance comes to 78,750,000 miles. Thus the modern figure of 93,000,000 miles for the distance from the earth globe to the sun is bracketed by the *Bhāgavatam* figures obtained using our two standard values for the length of a *yojana*.

If the planets do lie at great distances from us along the plane of Bhūmaṇḍala, then from our point of view the planets must always lie very close to the great circle on the celestial sphere corresponding to this plane. (We argued this for the sun in Section 3.d.) Now, is it true that the planets all tend to lie very close to some particular celestial great circle? The answer is yes. The orbits of all of the planets are observed to lie very close to the great circle, called the ecliptic, which is the geocentric orbit of the sun.

In Table 9 there is a list of the maximum distances of the planets from the ecliptic, according to modern astronomical data. These distances agree only roughly with the heights in Table 8, but they give the same order for the relative distance of the planets, and some are of the same order of magnitude.

(According to modern astronomy, Mercury should lie between Venus and Mars in this table because of the large inclination of its heliocentric orbit.)

One possible interpretation of Tables 8 and 9 is as follows: In accordance with the first hypothesis discussed in Section 3.d, the projection of the plane of Bhū-maṇḍala on the celestial sphere is the ecliptic. The *Bhāgavatam* is giving a qualitative description of how far the planets move from the ecliptic in the course of their orbits. In this description, the moon is higher than the sun because the sun always remains on the ecliptic whereas the moon moves away from it. Likewise, Venus is higher than the moon because it moves still further from the ecliptic.

One drawback of this interpretation is that the planets do not stay on one side of the ecliptic. In the course of their orbits they move equal distances on either side, following characteristic looping paths. This may seem to be in strong disagreement with the statements of the *Bhāgavatam*, which simply specify fixed heights for the planets. However, we have seen that Śrīla Prabhupāda has spoken of the disc of Bhū-maṇḍala as a system of globes floating in space, and we have also argued that this earth is a globe and was regarded as such in Vedic times. Furthermore, Śrīla Prabhupāda has said that planets belonging to different layers in the vertical direction can mix together in one layer. This may also seem contrary to the *Bhāgavatam*.

We propose that such apparent contradictions can be reconciled by the idea that the *Bhāgavatam* is using simple, three-dimensional imagery to describe a higher-dimensional situation that is directly experienced and understood by demigods, *ṛṣis*, and great yogis. In this case, we suggest that the image of perpendicular height above a plane provides a simple way to describe how the demigods view the actual, higher-dimensional situation: The height of a planet is an important higher-dimensional feature of that planet; this feature is reflected in the planet's visible motions away from the plane of the ecliptic and is described in simple terms in the Fifth Canto as height above the plane of Bhū-maṇḍala.

A final point concerning the seven planets is that the days of the week are named after these planets in both Europe and India. In Table 10 the names for the days of the week in English, Latin, and Sanskrit are given. These sets of names all refer to the seven planets in the order Sun, Moon, Mars, Mercury, Jupiter, Venus, and Saturn. Although this is not the order of the planets as given in Table 8, it does derive from Vedic astronomy.

In the *Sūrya-siddhānta* the planets are listed as follows in order of distance from the earth globe: Moon, Mercury, Venus, Sun, Mars, Jupiter, and Saturn. This list differs from the one in Table 8 since it refers to distance from the earth globe rather than distance from the plane of Bhū-maṇḍala. According to the *Sūrya-siddhānta*, the successive months of 30 days are ruled cyclically

TABLE 10

The Days of the Week

Planet	Day of the Week		
	SANSKRIT	ENGLISH	LATIN
Sun	Āditya-bara	Sunday	Solis dies
Moon	Soma-bara	Monday	Lunae dies
Mars	Maṅgala-bara	Tuesday	Martis dies
Mercury	Budha-bara	Wednesday	Mercurii dies
Jupiter	Bṛhaspati-bara	Thursday	Jovis dies
Venus	Śukra-bara	Friday	Veneris dies
Saturn	Śanaiścara-bara	Saturday	Saturni dies

The days of the week in Europe and India are named after the seven traditional planets.

by the planets in this order. (According to modern astronomy, the planets are sometimes aligned in the *Sūrya-siddhānta* order of distance from the earth, with the exception that Mercury and Venus must be switched.)

The successive days are ruled by the seven planets in such a way that the ruler of the first day of a month is always the same as the ruler of that month. If one places successive 7-day weeks next to successive 30-day months, one sees that if the first day of month 1 lines up with the first day of week 1, then the first day of month 2 lines up with the 3rd day of a week. Likewise, the first day of month 3 lines up with the 5th day of a week, and so on. This means that the days must be named after the planets according to the pattern shown in Table 11.

According to the dictionary, the English names for the days originated when the Latin names were translated into various Germanic dialects in about the third century A.D. Modern Western scholars trace the Latin names back to the Greeks, and as we might expect, they maintain that the Greeks originated these names. They also assert that Indian mathematical astronomy and astrology originated with the Greeks, and that the Sanskrit names for the days were translated from Greek at the time when this body of knowledge was imported into India. However, the history of this development is not known, and one can also argue that the system for assigning planetary names to divisions of the calendar is indigenous to India. After all, it is one thing for the Romans, who started their empire within the Greek sphere of influence, to have borrowed this system from the Greeks, and it is another thing for the long-established and highly conservative civilization of India to have done so.

TABLE 11

The Order of the Planetary Names of the Days of the Week

Order in Week	N	Remainder of (30N)/7	Order from Earth in the Sūrya-siddhānta
Sun	1	2	Moon
Moon	2	4	Mercury
Mars	3	6	Venus
Mercury	4	1	Sun
Jupiter	5	3	Mars
Venus	6	5	Jupiter
Saturn	7	7	Saturn

The rule given in the *Sūrya-siddhānta* is that the names of the 30-day months must match the names of their first days. The months are named cyclically in the order shown on the right, and the days must be named as shown on the left for the proper matching to occur.

4.b.1. Planetary Motion in the Bhāgavatam

In this subsection we will discuss the rates of orbital motion of the seven planets, as given in the *Śrīmad-Bhāgavatam*. In SB 5.21.3, 5.22.7, and 5.22.12 it is mentioned that the sun travels at three speeds: fast, slow, and moderate. These occur when the sun is in the south, in the north, and at the equator, respectively. These periods also correspond to the northern winter, when days are shorter than nights, the northern summer, when the opposite is true, and the time of the vernal and autumnal equinoxes. (Note that the word *equator* refers to the equinoxes, or times when day and night are equal.)

Some have interpreted these three *Bhāgavatam* verses to mean that days are shorter in the winter because in this season the sun moves across the sky faster during the day and slower during the night. However, the text of the *Bhāgavatam* does not say this, and at least two other interpretations are possible. The first of these assumes that the verses refer to the sun's daily motion. SB 5.23.3 compares the motion of the planets and stars around the polestar to yoked bulls walking around a central post threshing rice. Just as the bulls must walk faster the further they are from the post, so one can say that the sun's daily motion is faster the farther it is from the polestar. One can represent this mathematically by mapping the celestial sphere to a plane that is tangent to the north celestial pole.

The second interpretation assumes that the verses refer to the sun's yearly motion against the starry background. This assumption is supported by SB 5.22.12, which says that Venus shares the three speeds of the sun. Although this verse could refer to the daily motion of Venus, it is a fact that since Mercury, Venus, Mars, Jupiter, and Saturn are not generally visible during the day, one seldom (if ever) sees references to their daily motion. Also the verses following SB 5.22.12 all refer to the motions of these planets relative to the stars.

According to modern astronomy, it is in fact true that the sun moves faster along the ecliptic during the northern winter than it does during the northern summer. The heliocentric theory explains this as being due to the fact that the earth reaches perihelion, or its point of closest approach to the sun, just a few days after the winter solstice. At this time it is moving at its fastest rate in its orbit, and it is moving at its slowest rate exactly half a year later at aphelion. The *Sūrya-siddhānta* also gives calculations for the varying speed of the sun during the course of a year.

SB 5.21.4 states that the length of the day changes at a rate of one *ghaṭikā* per 30-day month during the period between the solstices. (A *ghaṭikā* is 24 minutes.) If we take this to mean one *ghaṭikā* in both the morning and the evening, then this rule is identical to a rule found in the *Vedāṅga-jyotiśa*, a short astronomical text said to be "one of the six *aṅgas* ['limbs'] of the *Vedas*" (VJ).

Some critics have scorned this rule as a crude approximation, and others have claimed that it works best at the latitude of Babylon, and is therefore Babylonian in origin. We programmed a computer to calculate the annual variation in the length of the day at various latitudes, using modern astronomy. We found that at the latitude of Delhi the *Bhāgavatam* rule works quite nicely, as long as one is about 20 days away from the solstices. The rule's average error in the length of the day, over a full year period, is about 6.6 minutes at Delhi. In contrast, the average error at Babylon is about 9.1 minutes, and the rule doesn't work well during any time of the year. One can argue that the rule is a practical approximation intended for use in northern India. Certainly it is simpler to apply than the modern calculations.

In SB 5.22.9 it is stated that the moon passes through each constellation in an entire day. These particular constellations are called *nakṣatras*, or lunar mansions; they are 27 in number, and are used to divide the ecliptic into 27 equal parts. In Section 6.e we discuss them in greater detail. In this verse the implication is that the moon completes one sidereal orbit (an orbit against the background of stars) in 27 days. This is an approximation. For comparison, the modern figure is 27.321, and the *Sūrya-siddhānta* gives 27.322.

This verse also states that the waxing and waning of the moon respectively creates day and night for the demigods, and night and day for the *pītas*, or forefathers. Since some demigods have a day of 360 earth days, this verse presumably refers specifically to the demigods living on the moon. The simplest interpretation is that these demigods live on one side of the moon (the side facing us) and the *pītas* live on the other side. However, SB 5.26.5 places Pitṛloka in the region between the Garbhodaka Ocean and the lower planetary systems. It would seem that some connection must exist between Pitṛloka and the moon, but more research will be needed to determine exactly what it is.

SB 5.22.8 also gives the orbital period of the moon, but it is hard to interpret. Here we will give a tentative interpretation that may need to be corrected in the future. The verse states that (1) the distance covered by the sun in one year is covered by the moon in two fortnights; (2) the distance covered by the sun in one month is covered by the moon in 2.25 days; and (3) the distance covered by the sun in a fortnight is covered by the moon in one day. From SB 5.22.9 we know that distance (3) must be 1⁄27 of a circle, or 13⅓°. This makes sense, since distance (2) must be 30°, or 2.25 times 13⅓°. This is because the sun travels 360° in a year and 30° in 1⁄12 of a year.

However, for (1) to be true, a fortnight must be 13.5 days, even though this period is normally 15 days. (The reason for this is that to go from the 13⅓° covered in one day to the 360° covered in two fortnights, we must multiply by 27, or 2 × 13.5.) This conclusion is backed up by the fact that the sun should certainly travel more than 13⅓° in 15 days.

If we accept the 13.5-day fortnight and divide 13⅓ by 13.5, we find that the sun travels .9876° per day. For comparison, the modern figure is .9856° per day. These rates of motion correspond to solar years of 364.5 days and the modern value of 365.257 days. The point we would like to make here is that the *Bhāgavatam*, with its 360-day year, may seem naive, but there is actually considerable sophistication behind its calculations. They are simply expressed in a way that seems unusual from the Western point of view.

SB 5.22.14 states that Mars crosses each sign of the zodiac in three fortnights if it "does not travel in a crooked way." This rate of motion is 30° in 45 days, or ⅔° per day. The crooked motion of Mars may be its retrograde motion, but it is hard to specify just when this begins and ends, since the path of Mars begins to curve before its motion actually reverses. Table 12 lists the percentage of time that Mars spends traveling at different speeds, calculated according to modern astronomy. From this table we can see that SB 5.22.14 is making a reasonable statement that must have been based on considerable knowledge of the movements of Mars.

SB 5.22.15 states that Jupiter travels through one sign of the zodiac in one Parivatsara. The names Saṁvatsara, Parivatsara, Iḍāvatsara, Anuvatsara, and Vatsara all refer to a year of 360 days (SB 5.22.7). This verse therefore indicates that Jupiter takes 4,320 days to complete one orbital revolution. The modern figure is 4,332.58 days, and differs by about .3 percent. Likewise, SB 5.22.16

TABLE 12

The Various Speeds of Mars

Degrees/Day	Percentage
below .000	10.3%
.000 to .200	4.7%
.200 to .400	7.1%
.400 to .500	5.9%
.500 to .550	4.5%
.550 to .600	6.5%
.600 to .650	11.3%
.650 to .700	23.6%
.700 to .750	26.1%
above .750	.0%

This table lists the percentage of time that Mars spends traveling at various speeds, calculated according to modern astronomy. The columns on the left indicate a number of speed intervals for the motion of Mars. (A speed below zero corresponds to retrograde motion.) The column on the right gives the percentage of time that Mars spends in these speed intervals. Mars spends most of its time at speeds approximating .667, which is given in the *Bhāgavatam*.

states that Saturn makes one orbital revolution in 30 Anuvatsaras, which means that Saturn takes 10,800 days to complete one revolution. Here the modern figure is 10,759.2 days and differs by about .38 percent. It is rather remarkable that the *Bhāgavatam* can express orbital periods with such accuracy using simple expressions such as "one sign per Parivatsara."

4.C. HIGHER-DIMENSIONAL TRAVEL IN THE VERTICAL DIRECTION

One aspect of our interpretation of the planetary distances in Table 8 is that the vertical dimension in Vedic cosmology is more than just a third coordinate axis, as understood in ordinary geometry. It also involves a higher-dimensional aspect that goes beyond the range of our senses. We can obtain one indication of this by considering the highest destination that one can reach within this universe by traveling in this vertical direction. This is the planetary system called Satyaloka, which is the abode of Brahmā, the secondary creator of the universe.

According to the *Bhāgavatam*, Satyaloka is located near the top of the universal globe, in the direction of the north celestial pole. Since the earth is located near the center of this globe, this means that Satyaloka is about 2 billion miles from the earth. A spaceship traveling at 500 miles per hour (a moderate speed for a jet plane) could cover 2 billion miles in 457 years, and thus it would seem that it might be feasible for human beings to reach Satyaloka using mechanical technology.

Yet in SB 5.1.21p we read the remarkable statement that Satyaloka "is situated many millions and billions of years away." Similarly, SB 1.9.29p states that "even attempting to reach the highest planet will take millions of years at a speed of millions of miles per hour." And SB 2.2.23p completely rules out the possibility of going beyond Svargaloka or Janaloka by "mechanical or materialistic activities, either gross or subtle."

SB 5.1.21 describes the abode of Brahmā as being "indescribable by the endeavor of mundane mind or words." In the terminology adopted in this book, this means that to describe Satyaloka adequately, we would have to make use of higher-dimensional concepts that cannot be grasped by our present minds and senses. At the very least, this implies that our ordinary concepts of space and time might break down when applied to this region of the universe.

An interesting indication of the form this breakdown might take is given in the following story from the *Bhāgavatam*:

> Taking his own daughter, Revatī, Kakudmī went to Lord Brahmā in Brahmaloka, which is transcendental to the three modes of material nature, and inquired about a husband for her. When Kakudmī arrived there, Lord Brahmā was engaged in hearing musical performances by the Gandharvas and had not a moment to talk with him. Therefore Kakudmī waited, and at the end of the musical performances . . . [he] submitted his long-standing desire.
>
> After hearing his words, Lord Brahmā, who is most powerful, laughed loudly and said to Kakudmī, "O King, all those whom you may have decided within the core of your heart to accept as your son-in-law have passed away in the course of time. Twenty-seven *catur-yugas* have already passed. Those upon whom you may have decided are now gone, and so are their sons, grandsons, and other descendants. You cannot even hear about their names" [SB 9.3.29–32].

Here we see that when one visits Satyaloka, one experiences a transformation of time reminiscent of the time dilation of Einstein's theory of relativity. King Kakudmī and his daughter were evidently advanced yogis who were able to travel to Satyaloka by nonmechanical means. Although for them the trip took only a short time, when they returned to the earth they found that millions of years had passed. We may then ask, Did the distance that they traveled seem like two billion miles to them? If so, then it might also be that from our perspective the distance was billions and billions of miles. Although this is merely a conjecture, it does indicate some of the things that are possible in a universe that is ultimately inconceivable by our mundane minds. (Note, by the way, that Revatī is the name of the star Zeta Piscium, which is used as the zero point for celestial longitudes in the *jyotiṣa śāstras*.)

Between the earth and Satyaloka there is a standard path traversed after death by transcendentalists and highly elevated persons. This is called the *uttarāyaṇa* path, and it is mentioned in the *Bhagavad-gītā* (8.24). A more detailed description of the various stages of this path is given in the *Vedānta-sūtra* commentary of Baladeva Vidyābhūṣaṇa:

> (1) Archis, the Deva of light, (2) Dinam, the Deva of day, (3) Śuklapaksam, the Deva of the Bright-fortnight, (4) Uttarāyanam, the Deva of the northern progress of the sun, (5) Samvatsaram, the Deva of the year, (6) Devalokam, the world of the Devas (the same as Vāyuloka, according to some), (7) Vāyu, the world of Vāyu, (8) Ādityam, the world of the sun, (9) Chandram, the world of the moon, (10) Vidyut, the world of lightning, (11) Varuṇam, the world of water, (12) Indram, the world of Indra, (13) Prajāpati, the world of Prajāpati, or of the four-faced Brahmā [VSB, p. 729].

Baladeva Vidyābhūṣaṇa comments that in this list, the various items refer not to landmarks on the path, but to various demigods who make arrangements for the passage of the soul (see BG 8.24p also). This indicates that higher-dimensional travel along the "vertical dimension" of the universe involves more than a simple ballistic trajectory of the kind followed by a rocket. It also involves the action of a hierarchy of beings, all of whom are inaccessible to our ordinary senses. The motion towards the north celestial pole is simply the three-dimensional aspect of this higher travel.

The descent of the Ganges River from the upper regions of the universe to the earth provides another interesting indication of the nature of travel along the vertical dimension in Vedic cosmology. According to the *Bhāgavatam*, the Ganges consists of water from the Kāraṇa Ocean that entered the upper portion of the universe through a hole kicked in the universal covering by Lord Vāmanadeva (SB 5.17.1). This water takes a thousand *yugas* to reach the planet Dhruvaloka, or the polestar, which is situated approximately 30 million miles above the sun. (Here the term *yuga* indicates a *divya-yuga* of 4,320,000 years.) Since the sun is situated vertically in the center of the universe (SB 5.20.43), this means that the Ganges covers a distance of some two billion miles in 4,320,000,000 years. Since this is a very slow rate of progress even for a very sluggish river, this may be another example of the transformation of time, and possibly of space, which occurs in the higher regions of the universe.

From Dhruvaloka the Ganges reaches the planets of the seven sages, and from there it is carried to the moon "through the spaceways of the demigods" in billions of celestial airplanes. From the moon it falls down (*nipatati*) to the top of Mount Meru, where it divides into four branches. Finally, one of these branches becomes the Ganges of India (SB 5.17.3–9).

Since the moon is continuously moving in its orbit, it is hard to see how the top of Mount Meru could always be directly underneath it in an ordinary geometric sense. It therefore seems that the descent of the Ganges from the moon to Mount Meru must involve physical principles that are presently unknown. Of course, as we have already pointed out, the final appearance of the Ganges in India also requires such principles, since we certainly do not see its descent from a higher region of the universe.

Thus our conclusion is that if we take the description of the descent of the Ganges seriously, then we must be prepared to view it in the context of principles that go beyond the framework of our familiar physical theories. We suggest that although these principles are not explicitly explained in the *Bhāgavatam* and other Vedic texts in Western terms, they are nonetheless employed in these works in a consistent way. One example of this is Śrīla Prabhupāda's statement in *Light of the Bhāgavata* that "one has to cross Mānasa Lake and then Sumeru Mountain, and only then can one trace out the orbit of the moon" (LB, p. 48). This statement is consistent with one idea that emerges from the story of the Ganges: In some higher dimensional sense, the route from the earth to the moon passes through the region of Mount Meru in Jambūdvīpa.

In SB 5.23.5 the celestial Ganges is identified with the Milky Way, and in SB 2.2.24 it is said that the Milky Way is a pathway that mystics follow through the heavens on their way to Satyaloka. It is interesting to note that similar ideas have traditionally been held in cultures around the world. Thus, both the Polynesians and various American Indian tribes maintained that the Milky Way is a pathway to heaven followed by the souls of the departed, and they also held that those souls who were not perfectly pure would eventually have to return to the earth (HM, p. 243).

The ancient Egyptians apparently regarded the Nile as an earthly continuation of the Milky Way (HM, p. 260), an idea they may have imported from an original homeland in India. (Śrīla Prabhupāda indicates in SB 2.7.22p that according to the *Mahābhārata*, the kings of ancient Egypt were driven there from India by Paraśurāma.)

The Chinese also had the idea that the Milky Way is a celestial river that descends to the earth. Their account is as follows: "The celestial river divides into two branches near the North Pole and goes from there to the South Pole. One of its arms passes by the lunar mansion Nan-teou (lambda Sagittarii), and the other by the lunar mansion Toung-tsing (Gemini). The river is the celestial water, flowing across the heavens and falling under the earth" (HM, p. 260).

4.D. THE ENVIRONS OF THE EARTH

According to the *Bhāgavatam*, the seven lower planetary systems have the same width and breadth as Bhū-maṇḍala, and they lie beneath Bhū-maṇḍala in successive strata separated by intervals of 80,000 miles (SB 5.24.7). Since the diameter of Bhū-maṇḍala is four billion miles, it follows that the complete system consisting of Bhū-maṇḍala and the seven lower worlds can be visualized as a relatively thin disc, comparable to a stack of eight circular sheets of paper. As with Bhū-maṇḍala, this means that the major part of the seven lower worlds lies many millions of miles away from us in what we regard as outer space. Also, the geometric projection of these lower worlds on the celestial sphere is nearly the same as the projection of Bhū-maṇḍala.

Traditionally, people in cultures throughout the world have spoken of subterranean realms, and the lower planetary systems described in Vedic literature also have a subterranean aspect. Thus the *Bhāgavatam* points out that the rays of the sun cannot reach the *bila-svarga* (SB 5.24.11), and in the *Mahābhārata* there are accounts of people traveling to these regions by entering tunnels leading down into the earth. Also, the astronomical *siddhāntas* place the lower worlds in the "concave strata of the earth."

The *Bhāgavatam*'s description of the dimensions of the lower worlds indeed suggests that these worlds constitute a subterranean stratum of Bhū-maṇḍala as a whole. According to this idea, this lower region does lie below our feet, and due to the vast extent of Bhū-maṇḍala, it also lies in outer space. Both of these locations, however, are simply three-dimensional projections (or aspects) of the actual, higher-dimensional position of the lower planetary systems. We could not reach Nāgaloka, for example, by tunneling into the earth using ordinary three-dimensional methods. But we could do so if our tunneling was accompanied by movement along a higher dimension.

Some 240,000 miles below the lowest of the seven lower worlds, Garbhodakaśāyī Viṣṇu lies on Ananta Śeṣa on the surface of the Garbhodaka Ocean (SB 5.25.1). As we have already noted, even Lord Brahmā was unable to see Garbhodakaśāyī Viṣṇu when he tried to trace out the origin of the lotus from which he himself had taken birth. It is therefore to be expected that this scene must lie beyond the range of ordinary human sense perception.

However, since the Garbhodaka Ocean is almost directly beneath the plane of Bhū-maṇḍala, one can imagine that all points on the celestial sphere south of the projection of Bhū-maṇḍala should correspond to this ocean. This amounts to a simple, three-dimensional visualization of an essentially higher-dimensional situation. Our thesis in this book is that in the Vedic civilization, the relationship between higher-dimensional realms and the visible firmament was visualized in this way. (In Section 3.b.3 we have noted the existence of ancient traditions that are consistent with this idea.)

According to the *Bhāgavatam*, as we move up from the earth's surface, we ultimately reach a point where clouds and wind are no longer found. This is the beginning of *antarikṣa*, or outer space. At an altitude of 800 miles above the base of *antarikṣa* are the abodes of the Rākṣasas, Yakṣas, and Piśācas, and at a still higher altitude are the realms of the Siddhas, Cāraṇas, and Vidyādharas (SB 5.24.4–6). For comparison, the highest clouds are about 50 miles up according to modern observations, and the American and Russian manned orbital flights range in altitude from about 100 to 900 miles above the earth (MSF, pp. 534–36). The various types of beings mentioned in these verses are known in Vedic literature for their great mystic powers, and they clearly operate on a level that is inaccessible to ordinary human senses. The Rākṣasas, of course, are particularly known for their inimical nature and their ability to create various kinds of illusions.

4.E. ECLIPSES

If we go 80,000 miles above the region of the Siddhas, Cāraṇas, and Vidyādharas, we come to the level of the planet called Rāhu. Some 80,000 miles above Rāhu we reach the level of the sun, which is said to lie between Bhūrloka and Bhuvarloka in the middle of *antarikṣa* (SB 5.20.43, 5.24.1). We note that these measurements account for only part of the distance from Bhū-maṇḍala to the sun, since this is given as 100,000 *yojanas* (or 800,000 miles) in SB 5.23.9p. In the Vedic literature it is often mentioned that Rāhu causes solar and lunar eclipses by passing in front of the sun or moon. To many people, this seems to blatantly contradict the modern explanation of eclipses, which holds that a solar eclipse is caused by the passage of the moon in front of the sun and a lunar eclipse is caused by the moon's passage through the earth's shadow. However, the actual situation is somewhat more complicated than this simple analysis assumes.

The reason for this is that the *Sūrya-siddhānta* presents an explanation of eclipses that agrees with the modern explanation but also brings Rāhu into the picture. This work explicitly assumes that eclipses are caused by the passage of the moon in front of the sun or into the earth's shadow. It describes calculations based on this model that make it possible to predict the occurrence of both lunar and solar eclipses and compute the degree to which the disc of the sun or moon will be obscured. At the same time, rules are also given for calculating the position of Rāhu and another, similar planet named Ketu. It turns out that either Rāhu or Ketu will always be lined up in the direction of any solar or lunar eclipse.

In Chapter 1 we have already described how the astronomical *siddhāntas* define the orbit of Rāhu, and a similar definition is given for Ketu. The positions assigned to Rāhu and Ketu correspond to the ascending and descending nodes of the moon—the points where the orbit of the moon (projected onto the celestial sphere) intersects the ecliptic, or the orbit of the sun. These nodal points

Figure 16. An Islamic picture drawn in A. D. 1272 showing the angel Shamhurash fighting the eclipse-dragon. This “dragon” is the Western version of the Vedic Rāhu and Ketu.

Figure 17. A picture of St. George and the Dragon painted by Raphaël in 1504, during the Rennaisance. This should be compared with Figure 16.

rotate around the ecliptic from east to west, with a period of about 18.6 years. One of them must always point in the direction of an eclipse, since the moon can pass in front of the sun or into the earth's shadow only if the sun, moon, and earth lie on a straight line. Thus, by placing Rāhu and Ketu at the nodal points of the moon, the *Sūrya-siddhānta* conforms both with the modern theory of eclipses and the Vedic explanation involving Rāhu and Ketu.

One objection that may be raised to the explanation given in the *Sūrya-siddhānta* is that it contradicts the Vedic statement that the moon is higher than the sun. However, we have seen that this statement refers to the height of the moon above the plane of Bhū-maṇḍala, and not the distance along the line of sight from the earth globe to the moon.

Another objection one might raise is that the explanation in the *Sūrya-siddhānta* seems to be a cheap compromise between the Vedic account of eclipses (which many will regard as mythological) and the modern account (which many will regard as an import into India from the Greeks). It is true that Rāhu and Ketu seem to play a rather superfluous role in the eclipse calculations given in the *Sūrya-siddhānta*. However, there are reasons for supposing that these planets do not appear in these calculations as a mere decoration.

The principal reason for this is that the positions of Rāhu and Ketu play an important role in astrology. This means that astrologers need some system of calculation that will tell them where Rāhu and Ketu are at any given time. We have argued in Chapter 1 that astrology has traditionally played an important role in Vedic culture. From this it follows that some methods for calculating the positions of Rāhu and Ketu have traditionally been required in Vedic society. Since we have no evidence that any other method of calculating these positions has ever been used, this can be taken as an indirect indication that the method used in the *Sūrya-siddhānta* has co-existed with the Vedic *śāstras* for a very long time.

Of course, by this argument we cannot conclude definitely that this particular method of calculation has always been used. But we can at least be sure that the Vedic society, with its emphasis on astrology and the astronomical timing of religious ceremonies, has always needed more than a mere qualitative story to account for eclipses and other astronomical phenomena.

In the West there is also a long tradition ascribing solar and lunar eclipses to the action of some celestial beings of a demonic nature. There these beings have also been associated with the nodes of the moon, and they are known as the head and tail of the dragon. The story of this eclipse-dragon may help give us some indication of how little we really know about history. Figure 16 is a medieval Islamic picture showing an angel severing the head of the eclipse-dragon. (This is reminiscent of the story of the decapitation of Rāhu by Lord Viṣṇu.) Figure 17 is a strikingly similar picture showing St. George, the patron saint of England, slaying a dragon. Unless this is a complete coincidence, it would seem that the story of the eclipse-dragon was somehow woven into the iconography of early Christianity without any indication of its significance being preserved. (St. George is said to have been born in Asia Minor in about A.D. 300, but there is apparently no information indicating how he came to be connected with a dragon (BD, p. 539).) Unfortunately, our knowledge of the ancient history of this story is practically nonexistent.

4.F. THE PRECESSION OF THE EQUINOXES

"Beings still greater than these have passed away—vast oceans have dried, mountains have been thrown down, the polar star displaced, the cords that bind the planets rent asunder, the whole earth deluged with flood—in such a world what relish can there be in fleeting enjoyments?" (Śrīla Bhaktivinoda Ṭhākura in VNB, p. 18)

The precession of the equinoxes is another astronomical phenomenon that seems to involve a contradiction between the Vedic description of the universe and the picture built up in recent times on the basis of observation. According to Vedic literature, the stars and planets execute a continuous daily rotation around a fixed axis that extends from Mount Meru through the polestar. This motion is generally described in such a way as to indicate that the polar axis of rotation is rigidly fixed. Thus we read that "all the planets and all the hundreds and thousands of stars revolve around the polestar, the planet of Mahārāja Dhruva, in their respective orbits, some higher and some lower. Fastened by the Supreme Personality of Godhead to the machine of material nature according to the results of their fruitive acts, they are driven around the polestar by the wind and will continue to be so until the end of creation" (SB 5.23.2–3).

In contrast, it is taught in modern astronomy that the polar axis of rotation is not fixed. It is supposed to rotate slowly about the axis of the ecliptic at a fixed angle of about 23.5°. (The axis of the ecliptic is the line perpendicular to the plane of the ecliptic.) One complete revolution is said to take place in about 25,770 years, and thus the amount of rotation occurring in one year is about 50.29" of arc. This slow shifting of the polar axis has two important consequences. One is that the position of the equinoxes will rotate slowly around the ecliptic at 50.29" per year. The equinoxes are the locations of the sun in its orbit at the times of the year when the day and night are of equal length. Since they correspond to the points of intersection between the ecliptic and the celestial equator, they will rotate at the same rate as the polar axis.

The other consequence is that the center of daily rotation of the heavens will move slowly in a large circle centered on the axis of the ecliptic. As time passes, the center of rotation will sometimes lie on a particular star, which will then be known as the polestar. At other times no prominent star will lie at this center. At present the polestar is Polaris, and it is said that in about 12,000 years it will be the star called Vega.

Although this appears to contradict the Vedic view, it turns out, as usual, that things are not as simple as they might seem at first. In the *Sūrya-siddhānta* there is a rule for calculating something that seems quite similar to the precession of the equinoxes. According to this rule, the position of the sun at the time of the equinox will slowly shift back and forth over a total angle of 54°. The time for one complete back-and-forth movement (covering 54° twice) is given as 7,200 years, and thus the movement occurs at a rate of 54" of arc per year (SS, pp. 29–30). A rule of this kind is said to describe trepidation of the equinoxes.

This rule seems rather artificial. It assumes that the motion makes an abrupt about face at the endpoints of the 54-degree interval, and thus one suspects that it may be intended simply as a rough approximation. However, it does predict the observed motion of the equinoxes over the period of two thousand years or so for which we have records of observations. And similar rules are given in the *jyotiṣa śāstras* that smoothly round off the motion at the endpoints of the interval of motion (BJS).

Another consideration here is that in SB 5.21.4–5 the times of the equinoxes and solstices are given relative to the zodiac. These timings are the same as in the Western zodiac. Thus the equinoxes occur in the beginning of the signs Meṣa and Tula, which correspond to the Western Aries and Libra. The Western zodiac moves with the precession of the equinoxes, and thus the equinoxes always occur when the sun enters Aries and Libra. However, the zodiac of the *jyotiṣa śāstras* has a fixed starting point at the star Zeta Piscium, and thus the position of the equinoxes in this zodiac should shift gradually with the passage of time.

According to modern calculations, the last time the equinoxes occurred at the beginning of Meṣa and Tula was about A.D. 650, and the time before that was some 25,800 years previously. However, according to the trepidation theory of the *Sūrya-siddhānta*, they would also have occurred at the beginning of Meṣa and Tula at the beginning of Kali-yuga.

In the *Siddhānta-śiromaṇi* we find the idea that the equinoxes precess through a complete circle. There it is stated that "the motion of the solstitial points spoken of by Muñjala and others is the same as the motion of the equinox: according to these authors its revolutions are 199,669 in a *kalpa*" (SSB1, p. 157). This comes to about one complete revolution in 21,636 years.

Since Śrīla Bhaktisiddhānta Sarasvatī studied both the *Siddhānta-śiromaṇi* and the *Sūrya-siddhānta*, and cited them in his commentary on the *Caitanya-caritāmṛta*, it would seem that the Vaiṣṇava tradition must include some conception of the precession of the equinoxes. Unfortunately, we have very little information on this topic, and thus further research will be needed to clarify the exact nature of this conception.

As a final point, we should note that according to the modern understanding of the precession of the equinoxes, the present polestar, Polaris, is now about 1° from the north celestial pole and will pass within 28' of it around the year 2100. One thousand years ago, Polaris was about 6.5° away from the pole, and a thousand years before that it was about 12° away. Since 6.5° is about 13 times the width of the full moon, it is hard to see how Polaris could have been regarded as the polestar prior to one thousand years ago.

Due to a lack of suitable bright stars, it would appear that there was no prominent polestar from a few centuries ago back to about 1200 B.C. At about this time the pole passed between Beta Ursae Minorus and Kappa Draconis, and one could have said, roughly speaking, that there were two polestars. To find a prominent, single polestar, however, one would have to go back to about 2600 B.C., when Alpha Draconis was situated at the pole.

We therefore ask, How can this be reconciled with the fact that the *Bhāgavatam* strongly states the existence of a single polestar? One would think there must have been a prominent polestar when the *Bhāgavatam* was written. Western scholars maintain that the *Bhāgavatam* was written about 1,000 years ago, at a time when there had been no notable polestar for thousands of years. Could there be some mistake in the scholarly dating of the *Bhāgavatam* or in the modern reconstruction of the past history of the polestar, or both? Further research may shed more light on this issue. However, Śrīla Prabhupāda has said, "Whether the Vedic calculations or the modern ones are better may remain a mystery for others, but as far as we are concerned, we accept the Vedic calculations to be correct" (SB 5.22.8p).

The Empirical Case for The Vedic World System

In Chapters 2 through 4 we have outlined an understanding of Fifth Canto cosmology based on the idea that higher-dimensional worlds exist in parallel with the three-dimensional continuum of our ordinary experience. Roughly speaking, by the "dimensionality" of space we mean the degree of access to other places that is possible at any given place for a conscious observer. This degree of access depends on the sensory arrangement of the observer, and thus we conceive of space as being relative to the consciousness of living beings.

Kṛṣṇa has full access to all locations at once, and thus for Kṛṣṇa, space has the highest possible dimensionality. It is therefore possible for Kṛṣṇa to appear in any location at will without having to travel, as we see in the story of Kṛṣṇa's appearance in the womb of Uttarā to save Mahārāja Parīkṣit. Another way of expressing this feature of Kṛṣṇa is to say that He is all-pervading.

For embodied beings in the material world, different levels of spatial access are possible, depending on their karmic status and corresponding sensory constitution. According to the Vedic literature, there are 400,000 species in this universe having humanlike form, and many of these have levels of sensory power superior to that of ordinary human beings of our modern civilization. These include, for example, the Kiṁpuruṣas, who are endowed with "mystic powers by which one can disappear immediately from another's vision and appear again in a different form" (SB 4.18.20).

These humanlike species all have their countries and dwelling places, even though these may not be visible or accessible to us. Indeed, our thesis is that many regions of the earth, or Bhū-maṇḍala, are not accessible to ordinary human senses. These regions are actually part of our immediate environment, but we can reach them only through higher-dimensional travel.

5.A. UNIDENTIFIED FLYING OBJECTS

In this subsection we will discuss some modern empirical evidence suggesting that we are part of a larger world of humanlike beings that is largely inaccessible to our senses and that may involve higher-dimensional inhabited realms. Before we begin, we should emphasize that all empirical evidence is faulty, since it is subject to the four defects of sensory imperfection, mistakes, illusion, and the tendency to cheat. This is particularly true of empirical evidence regarding phenomena that cannot be readily controlled or subjected to systematic experimentation. It is even more true of the evidence we shall consider here, which may involve the independent actions of living beings possessing human or superhuman powers. Evidence of this kind will tend to be controversial no matter how strong it is, since it contradicts fundamental assumptions lying at the root of modern Western civilization. Unfortunately, such evidence will also tend to be imperfect and fragmentary, since we are unable to control the phenomena involved and there is a natural tendency for people to suppress reports of these phenomena.

Thus far in this book, we have presented arguments that are intended to show that Vedic cosmology might be true. These arguments can be divided into two categories: (1) explanations that clarify Vedic cosmological ideas and hopefully make them more plausible and understandable, and (2) refutations of objections to Vedic cosmology raised by modern scientific theories. (Chapters 6 and 7 and Appendix 2 contain additional material in this category.) Although these arguments may remove various objections to Vedic cosmology, they do not provide any direct empirical evidence indicating that Vedic cosmology is true. Of course, according to the *paramparā* system, Vedic cosmology should be accepted simply on the basis of śāstric authority. However, the doubt may arise that if Vedic cosmology really is true, then it would seem strange if no empirical evidence could be adduced that directly supports it.

We suggest that there is actually abundant evidence for the existence of realms of intelligent living beings operating almost entirely outside the range of our ordinary senses. This evidence is what we would expect to find if Vedic cosmology is true, and it is definitely not what we would expect to find on the basis of accepted scientific paradigms. It can therefore be interpreted as giving support for the Vedic world view, although it does not refer directly to the structure of Bhū-maṇḍala and other features of Vedic cosmography.

This evidence falls into three broad categories: (1) folklore and traditional world views, (2) psychical phenomena, and (3) the evidence regarding unidentified flying objects, or UFOs. Each of these categories provides direct testimony indicating that interactions have occurred between human beings and other intelligent beings possessing paranormal or superhuman powers. In this chapter we will discuss category three, although, as we will see, these categories are interrelated and show considerable overlap.

There is extensive documentation on the subject of UFOs, which is largely generated by three groups of people: empirical investigators, debunkers, and

UFO cultists. One prominent characteristic of this field of study is that the evidence tends to generate strong emotions, both positive and negative, in the people involved. This makes objective discussion of the evidence difficult. Nonetheless, the UFO evidence can be potentially useful in helping people understand the overall validity of the Vedic world view, and therefore we will briefly consider it here.

We will begin by considering two examples of sightings of unidentified flying objects. The first sighting took place during the evening of July 14, 1952. Second Officer William Nash was at the controls of a Pan American DC-4 flying at 8,000 feet in the vicinity of Norfolk, Virginia, and Third Officer William Fortenberry was acting as copilot. It is described that the night was clear, with unlimited visibility, and the lights of Newport News could be seen out of the port window.

> Shortly after 8:00 P.M. (EST), both men spotted a reddish glow off in the distance, apparently east of Newport News. As the glow resolved itself into six bright points, it became obvious that the objects were approaching at a very high speed. Within seconds, the objects could be clearly recognized as reddish, glowing discs, as they streaked under the airliner. Then, abruptly, the entire group flipped on edge, made a sharp-angled turn, and reversed direction. As this was happening, the procession of six discs was joined by two more identical objects coming from under the plane, and all eight blinked out, back on again, and then off for good, while heading westward north of Newport News [RS, p. 138].

Both Nash and his copilot had been military pilots, trained in observing and identifying aircraft. They maintained that the discs looked solid, with sharp, well-defined edges. Based on their observations of the flight paths of the discs, they estimated that they had been traveling at least 12,000 miles per hour.

A closer sighting of what seemed to be a strange flying machine was reported by several witnesses near Exeter, New Hampshire, during the early-morning hours of September 3, 1965 (RS, pp. 176–78). At 1:30 A.M. Police Officer Eugene Bertrand investigated a parked car and found a distraught woman who claimed that her car had been followed for some 12 miles by a spaceship with red lights. Bertrand rejected this story, but was soon summoned back to his police station to investigate a similar story by 18-year-old Norman Muscarello. The teenager had burst into the station at 2:24 A.M. "in a state of near shock." He stated that while he was hitchhiking along route 150, a glowing object with pulsating red lights suddenly came floating across a nearby field in his direction. He said that the object was as big as a house and that it was completely silent as it moved toward him. After he dove for cover, the object backed away, and he flagged down a car, which took him to the police station.

Bertrand and Muscarello returned to the scene, and at about 3 A.M. both saw the object rise silently from behind two seventy-foot-tall pine trees. As Bertrand later described it, it was a "large, dark, solid object as big as a house. . . . It seemed compressed, as if it were round or egg-shaped, with definitely no protrusions like wings, rudder, or stabilizer" (RS, p. 177). The object had a row of five blinding red lights that blinked cyclically, casting a blood-red glow over the field and a nearby farmhouse. As nearby horses kicked in their stalls and dogs howled, it floated about two hundred feet off the ground with a fluttering motion, like a falling leaf.

This testimony was confirmed by officer David Hunt, who arrived on the scene in time to observe the object for five or six minutes as it departed in the direction of Hampton. The police also received a phone call from an excited man in Hampton, who reported seeing the object.

Here we will briefly touch upon some of the interpretations that have been proposed for such sightings, but we will not try to resolve the many controversial issues they involve. Broadly speaking, these sightings have been interpreted as involving (1) illusions or hoaxes, (2) secret military vehicles, (3) alien spaceships from other planets, and (4) vehicles piloted by beings from higher-dimensional realms. Without going into great detail, we would evaluate these interpretations as follows.

There are, of course, many instances in which sightings of strange phenomena turn out to be illusions or even deliberate frauds. However, there also seem to be many reports—such as the two we have summarized here—that are not amenable to this interpretation. If we dismiss either of these reports as the result of illusion or fraud, then it would seem that we must cast grave doubt on the reliability of human testimony in general. Let us therefore consider what consequences follow if we give at least as much credence to human testimony as is customarily done in courts of law.

The hypothesis of secret military vehicles may explain some sightings, but it seems doubtful that it can account for all of them. For example, if the flying discs seen by Nash and Fortenberry in 1952 were the product of human military technology, then one might ask why no technology of an even remotely similar nature had been used by any nation during World War II, only seven years before. Of course, it might be argued that technology had advanced by leaps and bounds in the post-war period, as shown by the example of computers. But this does not seem to be true of aerospace technology. For example, in the 1980s the space shuttle is being propelled into earth orbit by dangerous, unreliable solid-fuel booster rockets quite similar to the rockets used by the ancient Chinese, and atmospheric flight still depends on conventional propellers and jet engines.

We should also point out that more is involved here than mere technology; many sightings seem to involve phenomena that are incomprehensible in terms of the known principles of physics. Secret military developments certainly take place, but we know of no example in which fundamental scientific advances were made that were unknown to civilian scientists. For example, the basic scientific principles underlying the atomic bomb were well known to European scientists prior to World War II, and the Manhattan Project was devoted to routine but expensive engineering developments. It is difficult to see how government scientists working under conditions of secrecy could make spectacular advances in fundamental physics that remain inconceivable to scientists in the world at large.

The hypothesis of aliens from other planets also has its drawbacks, when presented in conventional form. Let us examine this hypothesis from the perspective of modern science. According to modern scientific thinking, the other planets of our solar system are devoid of life. Many scientists think that intelligent life may have developed on planets circling other stars, but they believe that this could happen only by a process of evolution similar to the process that has produced life on the earth. It is therefore important to note that prominent evolutionists have ruled out this possibility. These evolutionists point out that many random events are involved in the production of human beings, and the chance that something even remotely comparable to ourselves could evolve independently on another planet is essentially zero.

The evolutionist George Gaylord Simpson has raised this question in the following form: "Even in planetary histories different from ours might not some quite different and yet comparably intelligent beings—humanoids in a broader sense-have evolved?" (GS, p. 268) His answer is that the essential nonrepeatability of evolution makes this extremely unlikely. A similar conclusion was reached by the evolutionist Theodosius Dobzhansky (TD).

In fact, we agree with the analysis of Simpson and Dobzhansky, and we would go further by noting that, according to their reasoning, the probability is nearly zero that evolutionary processes would produce humans on the earth. By a probability of nearly zero, we mean a probability of the form 1 out of 10 to the power N, where N is a number in the hundreds or thousands. If an event occurs on one planet with this probability, then the probability that it will occur independently on two planets out of a billion possible planets is about 1 out of 10 to the power 2N-18. (Here we are assuming the existence of one billion planets suitable for life, and the 18 is the log of one billion squared.) In short, it seems highly unlikely that the evolution of matter would produce builders of flying machines on the earth, and far less likely that it would do this independently on other planets. (For a detailed discussion of the low probabilities associated with the evolution of advanced life forms, see the book *Mechanistic and Nonmechanistic Science* (MN), by the author.)

Of course, we can depart from the scientific hypothesis of extraterrestrial aliens by proposing, say, that one superintelligent being (i.e., Brahmā) may have created humanoids on other planets. However, we are still confronted by the fact that, according to modern astronomy, the nearest star is several light-years away, and most stars in this galaxy are hundreds or thousands of light-years away. Given the limitations imposed by the known laws of physics, a vehicle traveling between the nearest star and the earth would take several years at the very least to make the trip.

Many thousands of sightings of unidentifiable flying vehicles have been reported in the period following World War II, and practically all have involved brief encounters followed by no significant developments. Since it is inconvenient to make many journeys, each of which lasts for years, these observations suggest that either (1) the aliens have established local residences or (2) they are able to travel faster than the speed of light. Our point at this stage in the argument is that in making necessary modifications of the scientific extraterrestrial-alien hypothesis, we have brought it closer, step by step, to the Vedic world view. According to Vedic cosmology, there are 400,000 created humanoid life-forms in the universe. Many are locally based (such as the Yakṣas and Vidyādharas), and many are capable of unusual modes of travel (such as travel at the speed of the mind).

Another aspect of the UFO phenomenon is what could be called its psychic component. UFO sightings are frequently accompanied by telepathic impressions that observers tend to interpret as communications transmitted by UFO occupants. Psychic healings are reported in connection with UFOs, and UFO encounters are often followed by the appearance of typical psychical phenomena. Here is one case that illustrates some of these features (JV, pp. 173–76):

On November 1, 1968, a French medical doctor was awakened by calls from his 14-month-old baby shortly before 4:00 A.M. On opening a window, he saw two hovering disc-shaped objects that were silvery-white on top and bright red beneath. After moving closer for some time, the two discs merged into a single disc, which directed a beam of white light at the doctor's house. The disc then vanished with a sort of explosion, leaving a cloud that dissipated slowly.

The doctor testified that he had received a serious leg injury while chopping wood three days before. After the departure of the mysterious object(s), the swelling and pain from this injury suddenly vanished, and during subsequent days he also noticed the disappearance of all the chronic after-effects of the injuries he had received in the Algerian war.

During a two-year period following this incident, there was no recurrence of symptoms associated with either the war injuries or the leg wound. However, strange paranormal phenomena began to take place around the doctor and his family. According to the French scientist Jacques Vallee, "Coincidences of a telepathic nature are frequently reported, and the doctor has allegedly, on at least one occasion, experienced levitation without being able to control it" (JV, p. 176). The doctor apparently did not experience such things prior to his UFO sighting.

Psychical phenomena are a standard feature of human societies in all times and places, and they are referred to almost continuously in the Vedic literature. In modern human societies there seems to be an almost inverse relationship between the development of mechanical technology and the development of various psychic powers. However, in the Vedic literature we read of beings, such as the Dānavas of *bila-svarga*, who possess both advanced mechanical technology and mystic *siddhis*, and who are apparently able to combine the two. The UFO phenomenon seems to involve something similar, and this is another reason for thinking that this phenomenon can be better understood in terms of Vedic cosmology than in terms of standard theories involving high technology and interstellar evolution.

In addition to sightings of UFOs from a distance, there are many reports of close encounters with UFO occupants. These beings are often reported to communicate directly by telepathic processes, and they are also said to be able to project illusions through some kind of hypnotic power. Here is a typical example of this kind of report (JV, pp. 191–92):

On November 17, 1971, at 9:30 P.M., a Brazilian man named Paulo Gaetano was driving back from a business trip, accompanied by his friend Elvio. Paulo informed his companion that the car was not pulling normally, but his companion reacted by saying that he was tired and wanted to sleep. The engine then stalled, and after pulling to the side of the road, Paulo saw some kind of craft about twelve feet away. Next, he later maintained, several small beings appeared, took him into the craft, and subjected him to some kind of medical examination, which included taking a blood sample from his arm. He could not recall how he and Elvio got back home.

For his part, Elvio did not remember seeing a strange craft, but only an ordinary bus following the car at a normal distance. He saw the car pull off to the side of the road, and he remembered finding Paulo on the ground behind the parked car. But he did not remember seeing Paulo get out of the car, and did not know what had happened to him. He took Paulo by bus to the nearby town of Itaperuna, but he could not explain why they had abandoned the car. The police noticed the cut on Paulo's arm and later found the car parked on the highway.

Of course, there is a natural temptation to dismiss stories such as this as crazy nonsense. However, there are evidently many cases in which events of this kind are reported (including many that do not involve the questionable procedure of hypnotic regression). One possible explanation is that these stories involve delusions caused by some kind of mental disorder. However, there is psychiatric testimony indicating that common forms of nervous and mental disease do not involve delusions of seeing UFOs. For example, the psychiatrist Berthold Schwarz has stated,

> In thirteen years of private practice . . . I have never noted symptoms related to UFOs. A similar finding was confirmed on questioning Theodore A. Anderson, M.D., a senior psychiatrist, and Henry A. Davidson, M.D., then Medical Director of the Essex County Overbrook Hospital. Dr. Davidson recalled no patients with gross UFO symptoms out of three thousand in-patients, nor among all those presented to the staff while he was superintendent; nor of the thirty thousand patients who have been hospitalized since the turn of the century [ET, pp. 23–24].

It is possible that UFO close-encounter cases may involve the action of beings endowed with Vedic mystic *siddhis*. We do not wish to insist on this point, but we note that such a state of affairs would be consistent with Vedic cosmology. Śrīla Prabhupāda describes the *vaśitā siddhi* as follows:

> By this perfection one can bring anyone under his control. This is a kind of hypnotism which is almost irresistible. Sometimes it is found that a *yogī* who may have attained a little perfection in this *vaśitā* mystic power comes out among the people and speaks all sorts of nonsense, controls their minds, exploits them, takes their money, and then goes away [NOD, p. 12].

The story of Paulo and Elvio clearly involves some kind of illusion (either of the bus or of the strange craft). We should also note that many people reporting close encounters with UFOs maintain that the UFO occupants overcame their will with some kind of telepathic power.

The appearance of humanoid beings in UFO reports enables us to strengthen our remarks concerning the theory of evolution. The literature on UFOs is filled with reports of a wide variety of humanlike beings. These beings often exhibit recognizable emotions, and sometimes are said to communicate various philosophical teachings. If such beings actually exist, then it is very hard to see how they could have arisen by evolution, either on this planet or elsewhere. Paleoanthropology has no place for them on the earth, and the probability that beings so similar to ourselves would evolve independently on another planet is certainly infinitesimal. They fit consistently into the Vedic world view, but their existence is strongly incompatible with the theory of evolution.

Our final topic in this section is the tendency of UFO phenomena to abruptly appear and disappear from the viewpoint of human observers and their electronic instruments. Here are two cases illustrating this. The first case involved Air Force observations of a UFO in the south-central U.S. on July 17, 1957, and was summarized in the journal *Astronautics and Aeronautics*, as follows:

> An Air Force RB-47, equipped with electronic countermeasures (ECM) gear and manned by six officers, was followed by an unidentified object for a distance of well over 700 mi. and for a time period of 1.5 hr., as it flew from Mississippi, through Louisiana and Texas and into Oklahoma. The object was, at various times, seen visually by the cockpit crew as an intensely luminous light, followed by ground-radar and detected on ECM monitoring gear aboard the RB-47. Of special interest in this case are several instances of simultaneous appearances and disappearances on all three of these physically distinct "channels," and rapidity of maneuvers beyond the prior experience of the crew [AAA, p. 66].

One of the disappearances of the object occurred as the RB-47 was about to fly over it. The pilot remarked that it seemed to blink out visually and simultaneously disappear from the scope of ECM monitor #2 (an electronic surveillance device). At the same time it disappeared from radar scopes at ADC site Utah (a radar station on the ground). Moments later the object blinked on again visually, and simultaneously appeared on the ECM monitor and ground radar.

Abrupt appearances and disappearances of this kind are reported in many UFO accounts (including the Nash and Fortenberry sighting, with which we began this section). Although one might propose that invisibility was being produced through techniques involving known physical laws, this behavior of UFOs has suggested to many observers that they are illusions or projections of some kind, rather than physical objects. This is also suggested by the ability of these entities to accelerate abruptly to remarkable speeds without generating noticeable sonic booms. Of course, the hypothesis of illusion raises the question of how radar-reflecting illusions exhibiting intelligent behavior are generated.

The idea of illusion is also suggested by our second case, which took place at Nouatre, Indre-et-Loire, France, on September 30, 1954. At about 4:30 P.M. Georges Gatey, the head of a team of construction workers, encountered a strange-looking man standing in front of a large shining dome that floated about three feet above the ground. Our concern here is with the way in which these odd apparitions disappeared:

> Suddenly the strange man vanished, and I couldn't explain how he did, since he did not disappear from my field of vision by walking away, but vanished like an image one erases suddenly.
>
> Then I heard a strong whistling sound, which drowned the noise of our excavators; the saucer rose by successive jerks, in a vertical direction, and then it too was erased in a sort of blue haze, as if by a miracle [VJ2, p. 68].

Mr. Gatey, a pragmatic war veteran, maintained he was not used to flights of fancy, and his story was corroborated by several of the construction workers. Although such stories seem bizarre, they are not uncommon, and they are consistent with the more prosaic long-distance sightings reported by pilots and military personnel. They are also consistent with the mystic powers attributed to the Kiṁpuruṣas and other intelligent beings described in the Vedic literature.

5.B. THE LINK WITH TRADITIONAL LORE

When the reports of UFOs are surveyed broadly, they are seen to resemble stories from traditional folklore that have been recounted in cultures all over the world since time immemorial. Jacques Vallee illustrates this point with the following story from ninth-century France:

> One day, among other instances, it chanced at Lyons that three men and a woman were seen descending from these aerial ships. The entire city gathered about them, crying out they were magicians and were sent by Grimaldus, Duke of Beneventum, Charlemagne's enemy, to destroy the French harvests. In vain the four innocents sought to vindicate themselves by saying that they were their own country-folk, and had been carried away a short time since by miraculous men who had shown them unheard of marvels, and had desired to give them an account of what they had seen. The frenzied populace . . . were on the point of casting them into the fire, when the worthy Agobard, Bishop of Lyons, . . . having heard the accusations of the people and the defense of the accused, gravely pronounced that both one and the other were false [JV, p. 19].

The story refers to the "miraculous men" as sylphs, a class of beings thought by Paracelsus to inhabit the earth's atmosphere and to have the power of appearing or disappearing at will before humans.

In medieval folklore, such beings were thought to coexist with ordinary humans in this world and to inhabit invisible abodes, sometimes associated with lakes, mountains, or subterranean regions (EW). They were thought to interact with people in ways that were sometimes beneficial, sometimes sinister, and sometimes mischievous or trivial. Similar patterns of interaction are to be seen in the UFO literature.

According to the Vedic literature, interactions of this kind occur between humans and a variety of near-human beings, including Yakṣas, Kiṁpuruṣas, Rākṣasas, Vidyādharas, and Gandharvas. These beings occupy Bhū-maṇḍala, the lower planetary systems, and the upper system of Bhuvarloka. They are to be distinguished from the demigods and *ṛṣis* of Svargaloka and the higher planetary systems ranging up to Brahmaloka.

Such beings are frequently described in the Vedic literature as possessing aerial vehicles called *vimānas*. This is illustrated in the story of Śālva from *Śrīmad-Bhāgavatam*. There it is described that a king named Śālva engaged in severe austerities to please Lord Śiva and thereby obtain an airplane that could be used to attack Kṛṣṇa's city of Dvārakā. Lord Śiva granted the benediction and arranged for the airplane to be manufactured by the demon Maya Dānava, an inhabitant of the lower planetary system of Talātala in *bila-svarga*. The airplane is described as follows:

> But still the airplane occupied by Śālva was very mysterious. It was so extraordinary that sometimes many airplanes would appear to be in the sky, and sometimes there were apparently none. Sometimes the plane was visible and sometimes not visible, and the warriors of the Yadu dynasty were puzzled about the whereabouts of the peculiar airplane. Sometimes they would see the airplane on the ground, sometimes flying in the sky, sometimes resting on the peak of a hill, and sometimes floating on the water. The wonderful airplane flew in the sky like a whirling firebrand—it was not steady even for a moment [KB, p. 649].

We can compare the appearance and disappearance of Śālva's airplane with the "blinking on and off" of the UFO observed by the crew of the RB-47. The observers on the RB-47 also noted that their UFO sometimes generated two signals with different bearings on their electronic monitoring equipment.

We have argued that the domain of Maya Dānava can be reached only by higher-dimensional travel, and we suggest that even today, people of this earth may be interacting with beings originating from higher-dimensional regions of the universe. In Vedic times, people in general could directly see such phenomena as Śālva's airplane. But they presumably had little direct access to Maya Dānava's abode and could learn of the existence of such places only through hearing from higher authority. It can be suggested that we might be in a similar situation today.

In the *Bhāgavatam* it is described that the inhabitants of Maya Dānava's abode have excellent material facilities, including cities with beautiful architecture and attractive gardens. There is no fear of the passage of time there because the distinction between day and night does not exist. The inhabitants are highly atheistic and materialistic. They are expert in various mystic powers and are free from disease and old age. However, they must all meet eventual death in accordance with the strict arrangement of the Supreme Personality of Godhead (SB 5.24.10–14).

The Vedic literature describes the universe as having a hierarchical organization, with a graded series of domains occupied by beings with different levels of consciousness. As we described in Section 4.a, these domains can be divided into the lower, middle, and upper worlds, whose inhabitants are characterized by the respective modes of ignorance, passion, and goodness. Beyond the material world lies the transcendental domain of the Supreme Personality of Godhead, which is characterized by pure goodness (*viśuddha-sattva*).

Given this hierarchical structure, one would expect interactions between humans and higher beings to be characterized by a variety of psychological modes, ranging from ignorance up to pure goodness. This seems to be the case, and it is interesting to note that cases of interaction on an apparently higher, sattvic level provide some of the best-attested evidence for the existence of higher beings and realms.

5.C. THE EVENTS AT FATIMA

An example of interaction with higher beings on a sattvic level is provided by the events that occurred in 1917 in Fatima, Portugal. These events centered on a series of revelations made to three children named Lucia, Francisco, and Jacinta by a divine personage whom they understood to be the Virgin Mary. The revelations occurred on the 13th of the month for six successive months in a natural amphitheater called the Cova da Iria, near the town of Fatima. Here we will not be concerned with the theological content of the revelations (which, as far as it goes, is compatible with the philosophy of Kṛṣṇa consciousness), but will focus on the evidence they provide for the existence of higher-dimensional realms.

The revelations were made in the presence of the three children and a large throng of onlookers, which increased greatly from month to month as news spread. The actual visions of the beautiful divine personage could be seen only by the three children, and so our knowledge of these visions is limited to their testimony. However, during the revelations there occurred related phenomena that were witnessed by large numbers of people.

These phenomena included the appearance of a glowing, globe-shaped vehicle and the occurrence of a shower of rose petals that vanished upon touching the ground. One witness, Mgr. J. Quaresma, described the appearance of the globe on July 13, 1917, as follows:

> To my surprise, I see clearly and distinctly a globe of light advancing from east to west, gliding slowly and majestically through the air. . . . Suddenly the globe with its extraordinary light vanished, but near us a little girl of about ten continues to cry joyfully, "I still see it! I still see it! Now it is going down!" [FJ, p. 46].

He reports that after the events,

> My friend, full of enthusiasm, went from group to group . . . asking people what they had seen. The persons asked came from the most varied social classes and all unanimously affirmed the reality of the phenomena which we ourselves had observed [FJ, p. 47].

During one of the revelations, the child Lucia had requested that a miracle be shown so that people who could not see the divine lady would believe in the reality of what was happening. She was told that this would occur on the 13th of October, and she immediately communicated this to others.

On this date it is estimated that a crowd of some 70,000 people congregated in the vicinity of the Cova da Iria in anticipation of the predicted miracle. The day was overcast and rainy, and the crowd huddled under umbrellas in the midst of a sea of mud. Suddenly, the clouds parted, and an astonishing solar display began to unfold. We will describe this in the words of some of the witnesses.

Dr. Formigao, a professor at the seminary at Santarem:

> As if like a bolt from the blue, the clouds were wrenched apart, and the sun at its zenith appeared in all its splendour. It began to revolve vertiginously on its axis, like the most magnificent firewheel that could be imagined, taking on all the colours of the rainbow and sending forth multi-coloured flashes of light, producing the most astounding effect. This sublime and incomparable spectacle, which was repeated three distinct times, lasted for about ten minutes. The immense multitude, overcome by the evidence of such a tremendous prodigy, threw themselves on their knees [FJ, p. 63].

Dr. Joseph Garrett, Professor of Natural Sciences at Coimbra University:

> The sun's disc did not remain immobile. This was not the sparkling of a heavenly body, for it spun round on itself in a mad whirl, when suddenly a clamour was heard from all the people. The sun, whirling, seemed to loosen itself from the firmament and advance threateningly upon the earth as if to crush us with its huge fiery weight. The sensation during those moments was terrible [FJ, p. 62].

Similar testimony was given by large numbers of people, both from the crowd at the Cova da Iria and from a surrounding area measuring about 20 by 30 miles. The presence of confirming witnesses over such a large area suggests that the phenomena cannot be explained as the result of crowd hysteria. The absence of reports from a wider area and the complete absence of reports from scientific observatories suggest that the phenomena were local to the region of Fatima. It would seem either that remarkable atmospheric phenomena were arranged by an intelligent agency at a time announced specifically in advance, or that coordinated hallucinations in thousands of people were similarly arranged at this time. By either interpretation, it is hard to fit these phenomena into the framework of modern science. They do, however, fit naturally into the multidimensional, hierarchical cosmos of the Vedic literature.

At this point we should briefly summarize the conclusions of this chapter. The Vedic literature maintains that we live in a hierarchically structured universe occupied by 400,000 species of humanlike form and some 8,000,000 nonhuman species. These living beings inhabit a graded system of worlds such as Bhū-maṇḍala, which possess variegated geographical features. The thesis of this book is that these worlds are literally real, even though they are almost entirely inaccessible to ordinary human senses. We have tried to relate this idea to modern mathematical thought by describing the universe as a multidimensional system.

We can ask, If this description of the universe is correct, then is there anything that humans could expect to observe that would tend to corroborate it? First of all, we could not expect to readily see the demigods, for human beings generally do not have the karmic qualifications required for this. (See Section 6.c.1.) This means that we also cannot expect to gain access to the regions where the demigods are active, such as the slopes of Mount Meru, since this would surely entail being able to see the demigods themselves.

We might expect to interact with beings lower than the demigods but slightly higher than ourselves in the cosmic hierarchy. However, since these beings are higher than ourselves, we could not expect to fully control these interactions. We might expect that they would have access to us but that we would not have access to them or their abodes.

The beings between humans and demigods range from powerful types of ghosts to demonic entities such as Yakṣas and Rākṣasas, and to more attractive beings such as Vidyādharas, Cāraṇas, Siddhas, and Gandharvas. Interactions with these various beings might take various forms, depending on their own interests and on our level of consciousness. These interactions are likely to involve mystic powers, since these beings are all well-endowed with such powers. They must involve higher-dimensional transformations, because the abodes of these beings are invisible to our senses. They may also involve remarkable flying machines, since such machines are often ascribed to these beings in the Vedic literature.

We propose that these ideas about what we might expect to see in the Vedic universe are consistent with the evidence provided by psychical phenomena, UFO reports, folklore, and reports of miracles. This broad body of evidence is consistent with the Vedic world picture. However, none of this evidence is compatible with modern science, and all of it is rejected by the scientific community. We suggest that the Vedic world view is broadly supported by empirical evidence, but this evidence can never be respectable until the Vedic world view itself is restored to a respectable status.

Modern Astrophysics And the Vedic Perspective

Thus far we have tried to show how Vedic cosmology relates to the overall picture of the universe that modern man has built up on the basis of ordinary sense perception. We have mainly dealt with fairly elementary features of this picture that have been part of human knowledge for centuries. In recent years, however, a highly sophisticated and complex science of astrophysics has grown up, which deals with many celestial phenomena in great technical detail. Many people will be inclined to argue that Vedic cosmology is no match for this new science, in which astronomy has joined forces with physics. They will say that it describes nature on a level of precision and detail that was never approached in ancient times, and it has established many new concepts that were undreamt-of by earlier thinkers. These dynamic developments stand in sharp contrast to the static Vedic world view and show that its many unverifiable ideas have simply been a hindrance to progress.

One answer to this challenge is that to appreciate Vedic cosmology, one must understand its underlying purpose and the basic tenets on which it is based. These are quite different from the motives and assumptions lying behind modern science, and thus it is not surprising that they should lead to radically different scientific and cultural developments. We can evaluate the relative merit of the Vedic and modern approaches to nature only if we take these fundamental goals and assumptions into account.

The picture of Vedic cosmology that emerges from this study is based on the fundamental principle that reality can be understood only partially and imperfectly through the endeavors of our limited mundane minds and senses. Thus Vedic descriptions of the universe have stressed the existence of higher realms of being, both material and spiritual. Since these realms could not be understood in detail by persons on the human level of consciousness, they were described only in general, qualitative terms. For those who wished to know of these realms more directly, processes of yoga were given, which enable a person to gradually elevate his consciousness to higher levels. Thus human endeavor was channeled in the direction of purification of the self and the development of our inner potential, rather than toward the exploitation of the material environment using our gross sensory equipment.

Observable natural phenomena, such as the motions of the planets, were studied and mathematically analyzed in the ancient Vedic culture. However, the object of the analysis was not to give a final, comprehensive description of nature. Rather, its purpose was to provide simple and practical methods of dealing with these phenomena in day-to-day life. The motions of the planets were studied for the purpose of making astrological forecasts and arranging for the proper timing of religious festivals and ceremonies. Thus observational and mathematical astronomy was used to fulfill needs related to higher aspects of reality that cannot be directly observed and measured. Since there was no question of creating a final, mathematically perfect theory of astronomy, no efforts were made toward this end, and mathematical models were kept as simple as possible, given the practical needs for which they were intended.

According to the Vedic world view, the higher material and spiritual realms are by no means devoid of life. Rather, they are populated by a hierarchy of superhuman beings, and their original source is understood to be a supreme sentient being. Given this perspective, it is natural to think that knowledge about the most important aspects of reality can be obtained only by communicating with higher beings, and ultimately by coming in direct contact with the Supreme Lord. Thus the Vedic culture is dominated by the idea of receiving knowledge from a chain of authorities who are passing it down from a higher source. This applies not only to spiritual knowledge, but also to material arts, including mathematical astronomy.

In contrast, modern science is based on the idea that nature can be fully understood using our present minds and senses. Its fundamental premise is that all aspects of reality can be mathematically described, and that all phenomena can be observed either directly or through their effects on other phenomena. This leads naturally to the idea that it is possible to create a final, complete mathematical theory of nature. If we examine the history of modern physics and astronomy, we see that these fields of study have been dominated by the drive to pry loose all the secrets of nature quickly and to create such an ultimate theory.

We can therefore argue that many of the differences between Vedic and modern cosmology are due to this fundamental difference in approach. Vedic cosmology does not exhibit the same elaborate mathematical development as modern cosmology because the Vedic world view provided no motive for undertaking such a development. On the other hand, modern cosmology is strictly limited to a three- or four-dimensional continuum because modern man lacks the sensory faculties for observing higher-dimensional aspects of the universe, and because modern science places great emphasis on quickly arriving at a complete world model based on available observations.

Modern cosmology may seem superior to its Vedic counterpart if we stick to the assumption that reality is limited to what ordinary human beings can perceive, using either their unaided senses or mechanical instruments. However, if the Vedic idea of higher realms of existence is even approximately correct, then it becomes clear that the modern scientific approach has caused us to focus our attention uselessly on relatively unimportant aspects of the universe. From this point of view, the technical sophistication of modern astrophysics appears more as an impediment to the attainment of knowledge than as an example of great scientific progress.

To a person acquainted with modern scientific ideas, the obvious reply to this argument is that the complex technical methods of modern astrophysics have revealed many features of nature that contradict the Vedic literature, and thus the Vedic world view is no longer relevant. However, it is possible that a theoretical description of nature could be developed that equals or surpasses modern astronomical science in technical sophistication but is also consistent with Vedic cosmology. Such a theory might take the form of a radically new conceptual framework that incorporates our current theoretical system as an approximation having a limited range of applicability.

The point can be made that modern cosmology not only contradicts the Vedic literature but also has its own internal contradictions. These contradictions are quite severe, and we briefly discuss some of them in Chapter 7. They indicate that some radical change will have to be made in modern theories to bring them into line with astronomical observations. It is perhaps reasonable to suggest that such a revision should also take into account the empirical evidence for higher-dimensional aspects of reality discussed in Chapter 5. Such a new theoretical system might well agree more closely with Vedic cosmology than the present system does.

Radical extensions of our theoretical perspective have taken place repeatedly in the history of science. A striking example of this is provided by the revolution in the science of physics that occurred in the twenties and thirties of this century. At the end of the nineteenth century, physicists were almost universally convinced that classical physics provided a final and complete theory of nature. However, a few years later, classical physics was replaced by a new theory, called quantum mechanics, which is based on fundamentally different principles.

The most interesting feature of this development is that classical physics turns out to be compatible with quantum mechanics in the domain of observation in which it was originally applied. The differences between the two theories become significant only in the new atomic domain opened up by the quantum theory. Likewise, our proposed new cosmology would agree with existing theories in its predictions of gross sensory observations, but it would open an entirely new world of higher-dimensional travel.

6.A. THE PRINCIPLE OF RELATIVITY AND PLANETARY MOTION

To construct such a new cosmology, there are a number of important topics that must be considered. One of these is the idea of relativity of motion. The watershed in the development of modern astronomy was crossed when Copernicus replaced the ancient geocentric model of the universe with a heliocentric model. Although the relative merit of the two models was initially debatable, the development of Newton's laws of motion seemed to give overwhelming support for the heliocentric model. This can be argued as follows: If the stars and planets are rotating around the earth once per day, then they should be subjected to tremendous centrifugal forces that will have to be counterbalanced in some way. Isn't it more reasonable to suppose that the earth, which is much smaller and more compact than the universe as a whole, is rotating on its axis? Likewise, isn't it more reasonable to suppose that the small earth is orbiting around the massive sun than to suppose that the sun is orbiting around the earth?

This objection can be partially answered by invoking the idea of relativity of motion. Consider two objects, A and B, that are approaching one another at a constant velocity. According to classical physics, there is no *physical* difference between saying that A is standing still and being approached by B and saying that B is standing still and being approached by A. Thus, as far as physics is concerned, no objection could be raised to either statement.

In classical physics this relativity of motion is not thought to apply to rotation. Imagine an axis running from the center of A through the center of B. Suppose that A is rotating with respect to B on this axis. According to classical physics, rotary motion generates centrifugal force, and thus the actual rate of rotation of A and B can be determined by measuring this force. Thus if A exhibits a certain amount of centrifugal force and B does not, the conclusion of classical physics must be that A is rotating and B is not.

However, the physicist Ernst Mach once made the following argument: Suppose that A and B are the only objects in the universe, and suppose that they are of equal mass. Then why should it be that A shows measurable evidence of rotation and not B? After all, if we say that A is rotating, then what is it rotating with respect to? If B is the only other object in the universe, then A could only be rotating with respect to B. But it could equally well be said that B is rotating with respect to A. Thus Mach concluded that neither A nor B would exhibit centrifugal force if they were the only objects in the universe. He proposed that centrifugal force is generated in one object due to the rotation relative to it of another, much larger object. Thus, Mach maintained that if A is rotating with respect to the rest of the universe, then one could equally well say that the universe was rotating with respect to A and thereby generating centrifugal forces in A. Mach's argument implies that there are no *physical* grounds for rejecting the statement that "A is standing still and the universe is rotating around it."

Here one might object that the rotation of the earth is directly indicated by the Foucault pendulum experiment and the evidence that the prevailing winds are affected by Coriolis forces. Also, the rotation of the earth around the sun is indicated by a number of minute but measurable effects, such as aberration of starlight and the parallax of some stars.

It turns out, however, that Mach's argument also disposes of these objections. For example, Mach would say that the rotation of the Foucault pendulum can be attributed to the rotation of the massive universe around the earth, just as well as to the rotation of the earth under the pendulum.

If this idea of relativity of motion is granted, one can then argue that the geocentric or heliocentric viewpoints stand on the same footing physically, and we can choose one or the other, depending on what is convenient. In the case of the astronomical *siddhāntas*, we could argue that the geocentric viewpoint is simply the more practical of the two, since all computations must ultimately be expressed in geocentric terms. And if we intuitively prefer to think of large masses as stationary and small masses as moving, rather than the other way around, then we will prefer the heliocentric viewpoint.

When we turn to the cosmology of the *Bhāgavatam*, the situation becomes more complex. It is stated that the *pravaha* wind carries the celestial bodies around the polestar once per day. This can be seen from the viewpoint of relativity of motion in the following way: The *pravaha* wind is due to a kind of tenuous atmosphere that exists in the region of *antarikṣa*, or outer space. If we regard the earth as turning on its axis, then the stars are at rest in this stationary atmosphere. In contrast, if we regard this atmosphere as rotating along with the stars, then the stars are being carried by it, but they are still at rest in it. This brings to mind the analogy of the clouds and the wind that Śrīla Prabhupāda uses to illustrate the effects of *māyā*: Just as the clouds seem to be at rest in the wind that carries them, so people carried by the influence of *māyā* do not notice this influence.

The situation of Bhū-maṇḍala can be analyzed as follows: As we pointed out in Chapter 3, if Bhū-maṇḍala is located in the plane of the ecliptic, then Bhū-maṇḍala must also rotate daily with the *kāla-cakra*. The movement of the sun in Bhū-maṇḍala consists of one leftward revolution around Mount Sumeru per year, and both Bhū-maṇḍala, the sun, and the other planets are carried in one rightward rotation per day by the *pravaha* wind. Here, from the perspective of relative motion, one can regard the earth as rotating and the stars, *pravaha* atmosphere, and Bhū-maṇḍala as stationary. The sun is then seen to rotate with respect to Bhū-maṇḍala, being carried by its chariot. From the perspective that larger masses should be viewed as stationary, it is reasonable to regard the sun as moving and Bhū-maṇḍala as stationary, since Bhū-maṇḍala is much greater than the sun.

If we then take the covering shells of the universe into account and consider that the *pravaha* wind is blowing with respect to these fixed coverings, we obtain the following picture: It makes sense to suppose that the *pravaha* wind and the various celestial bodies are moving with respect to the universal coverings, since the coverings are more massive than the celestial bodies. Likewise, in this picture it also makes sense to suppose that the sun is moving with respect to Bhū-maṇḍala. This, of course, is the picture of celestial motion given in the *Bhāgavatam*.

As we mentioned in Chapter 3, the idea of relativity of motion is presented by Śukadeva Gosvāmī in his description of the motion of the sun. Mahārāja Pariksit asked him,

> My dear lord, you have already affirmed the truth that the supremely powerful sun-god travels around Dhruvaloka with both Dhruvaloka and Mount Sumeru on his right. Yet at the same time the sun-god faces the signs of the zodiac and keeps Sumeru and Dhruvaloka on his left. How can we reasonably accept that the sun-god proceeds with Sumeru and Dhruvaloka on both his left and right simultaneously? [SB 5.22.1]

Here the leftward and rightward movements are the yearly and daily revolutions of the sun about the earth. Śukadeva Gosvāmī replied to this question as follows:

> When a potter's wheel is moving and small ants located on that big wheel are moving with it, one can see that their motion is different from that of the wheel because they appear sometimes on one part of the wheel and sometimes on another [SB 5.22.2].

Śukadeva Gosvāmī explains that in this analogy the potter's wheel corresponds to the *kāla-cakra*, which carries the stars and signs of the zodiac with it. The ants correspond to the sun and other planets, which are moving leftward around the wheel while the wheel spins to the right. Thus the idea that motion can be seen differently from different relative perspectives is presented in the *Bhāgavatam*. We have discussed these points in some detail to show that Vedic cosmology should not be rejected on the basis of naive arguments regarding the relative motion of the earth, the sun, and the universe as a whole. To fully relate Vedic cosmology to the laws of motion of modern physics, it will be necessary to understand the bearing that structures such as Bhū-maṇḍala and the coverings of the universe have on our understanding of the principle of relativity. Since these structures involve higher-dimensional travel and transformations of time such as that seen in the story of King Kakudmī and Revatī, we do not think that this will be an easy task. But it may well be possible, and the resulting model will no doubt be even more surprising than the quantum theory was to the physicists of the early twentieth century.

We should also note that Einstein's theory of relativity is required in order to make sense of the heliocentric theory of the solar system. The history of this theory is that in the late 19th century, ether-drift experiments performed by physicists such as Michelson and Morley seemed to indicate that the earth is stationary relative to the ether. Since the ether was then conceived as a highly rigid medium, this seemed to indicate that the earth was stationary with respect to an absolute reference frame. Although many efforts were made to avoid this conclusion, this did not prove to be possible within the framework of classical physics.

The dilemma was resolved only with the introduction of Einstein's theory, which involved radical changes in physicists' concepts of space and time, and which has many strikingly counter-intuitive consequences. These include the famous twin paradox, in which a space traveler returns to earth from a year's journey at nearly the speed of light and discovers that many years have passed. It would take us too far afield to delve into these matters here, but we mention them as an indication that the issue of geocentric versus heliocentric cosmology is not as trivial as it might superficially seem to be.

6.B. GRAVITATION

The Newtonian theory of gravitation will play an important role in any attempt to harmonize modern physics and Vedic cosmology. This theory provides a uniform explanation of planetary motion that is tied conceptually to the heliocentric theory of the solar system. Quantitatively, it is highly accurate, and it has been confirmed by the experience people have gathered by launching artificial satellites and other vehicles into outer space. Since it provides an explanation for many details of planetary motion, many people will argue that it must be giving a correct account of the fundamental causes underlying planetary motion. However, even though this theory has been highly successful, it does have some shortcomings. These include the following:

(1) To this day, Newtonian theory cannot account for the long-term behavior

of the outer planets, namely Uranus, Neptune, and Pluto. One way to account for this is to posit the existence of an additional planet (or planets) that is influencing these planets. However, such a planet is thus far unknown.

(2) The story is often told that the French astronomer Leverrier predicted the position of the then unknown planet Neptune by gravitational calculations based on the orbit of Uranus, and that Galle in Berlin pointed his telescope in the indicated direction and found the planet right where Leverrier said it would be. This created an international sensation at the time. In addition, John Adams of England independently made calculations giving nearly the same prediction as Leverrier. However, further analysis quietly showed that "the planet Neptune is not the planet to which geometrical analysis had directed the telescope, and that its discovery by Galle must be regarded as a happy accident" (PL, p. 125). The discovery of Pluto involves a similar story (DR).

The erroneous stories of the discovery of Neptune and Pluto by gravitational calculation are still being repeated in various books and articles. This shows that the literature of modern astronomy is not fully reliable.

(3) In the 1870s Leverrier argued that discrepancies in the orbit of Mercury could be explained by the existence of a planet between Mercury and the sun. Such a planet was, in fact, repeatedly observed. It was called Vulcan, and Leverrier calculated an orbit for it, based on observations. Now, however, it is believed that this planet never existed and that the reported observations of it were all illusory. If this is so, then the derivation of an orbit from spurious observations suggests that considerable fudging was involved in Leverrier's calculations. On the other hand, if a planet-sized object did travel in Leverrier's orbit, then what became of it? (CR1, pp. 46–71)

(4) One of the most striking theoretical developments of 20th-century physics was Einstein's general theory of relativity, which accounted for the anomaly in Mercury's orbit. However, some have claimed that Newtonian theory can explain this if the sun's shape is sufficiently oblate (CR1, p. 28). And others have pointed out that there is an anomaly in the orbit of Venus that cannot be accounted for if Einstein's theory correctly accounts for the anomaly in Mercury's orbit (CR1, pp. 132–33).

(5) During Nov. 11–12, 1940, over 200 observers cooperated in studying the transit of Mercury across the sun. The transit began 36 seconds late and lasted 18 seconds less than it should have, according to gravitational calculations (CR2, p. 27). Transits of Galilean satellites across Jupiter also have been repeatedly reported to occur minutes from their calculated times (CR2, p. 79).

(6) Theories of planet formation based on Newtonian dynamics require that all planets should rotate on their axes in the same direction in which they rotate around the sun (i.e., counterclockwise as seen from the north celestial pole).

Recent radar measurements have shown that Venus revolves on its axis in a clockwise direction and always keeps one side facing the earth at times when Venus is closest to the earth. This is hard to explain, since tidal influences of the earth on Venus should be very weak. We should also note that pre-radar measurements showed that Venus rotates in a counterclockwise direction with a period of either 23 hours or 225 days. Recent measurements have also shown that the atmosphere of Venus has a clockwise rotation period of 5 days (CR2, pp. 302–4).

(7) The rings of the planet Saturn have many puzzling features, including the presence of many annular gaps. These are strange enough to provoke the following assessment:

> At the very least, resonance theory cannot account for the thousands of gaps—there are not nearly enough resonances. Indeed, some astronomers ask whether resonances can really explain any gaps. Sweeper moons might plow out some gaps, but the Voyager photographs do not reveal these postulated satellites. More ominously for celestial mechanics, the complex, dynamic nature of the rings seems beyond the power of Newtonian dynamics to explain and may require a whole new theoretical structure [CR2, p. 282].

(8) In May of 1976 the Laser Geodynamic Satellite was placed in an accurately determined orbit at an altitude of about 3,700 miles. The satellite was found to lose altitude at roughly ten times the rate attributable to aerodynamic drag and other known forces (CR2, p. 13).

(9) Small discrepancies in the orbital motion of the moon have led some investigators to propose that the gravitational constant G is slowly changing (CR1, pp. 260–64, 688).

(10) A team of researchers in Greenland has recently reported evidence for a small, non-Newtonian component in the force of gravity, and similar results have been reported by other investigators. It is interesting to note that the Greenland team includes physicists dedicated to new quantum mechanical theories of gravitation that make non-Newtonian predictions (DS).

The gravitational discrepancies in this list mostly involve small effects, but we include them to show that existing theories of gravitation are approximate descriptions of nature rather than exact accounts of how nature works. These examples also show how illusion and wishful thinking can play a role in making scientific theories seem more perfect than they actually are.

The underlying causes of gravitation have been a topic of controversy in the science of physics for a very long time. Newton himself stressed that his theory was only a numerical description of observable effects, and he deliberately made no hypotheses about underlying causes. He spoke of gravitation as "action at a distance," but the idea of a force acting mysteriously across empty space seemed abhorrent to Newton and other scientists, both in his day and the present. Thus the history of physics in the 18th and 19th centuries was marked by many attempts to explain gravitation through some kind of interaction of substances or particles moving through space. Unfortunately, all of these attempts were unsuccessful (RP, pp. 77–78).

In recent years Einstein's general theory of relativity has explained gravity as a bending of four-dimensional space-time. However, this theory has not been accepted as final by physicists, and attempts are now being made to formulate a quantum mechanical theory of gravitation. Since quantum mechanics is now accepted by physicists as the basis for understanding all atomic phenomena, such a theory is required to provide a consistent foundation for modern physics. Thus far, however, physicists have encountered insurmountable difficulties in their efforts to construct a quantum theory of gravitation, and the nature of gravity remains an open question.

Śrīla Prabhupāda has pointed out that, according to the Vedic understanding, planets float in outer space by the manipulation of air (SB 5.23.3p). He has rejected the idea of gravitation, calling it an imaginary law, but he has also said that the visible effects produced by the real causes of planetary motion can be called gravitation if one so desires. Since the issue of gravitation is so important, we should make a few observations about these statements.

First, when the Vedic scriptures speak of the planets being carried by the wind, we might think they are naively assuming that our atmosphere extends all the way to the planets. However, we have seen in Section 4.d that outer space, or *antarikṣa*, is said in the *Bhāgavatam* to begin a short distance above the earth at the upper limit of the clouds and ordinary winds. Thus the *pravaha* wind, which carries the planets, is of a different character than the winds of this earth. (The *Siddhānta-śiromaṇi* lists seven different types of winds, including *āvaha*, or atmosphere, and *pravaha*. See Appendix 1.)

Second, we should note that the Vedic literature also states that the planetary systems are supported by the Ananta Śeṣa expansion of Lord Viṣṇu. This can be reconciled with the statement that the planets float by manipulation of air if we suppose that the action of Ananta Śeṣa is the fundamental cause of planetary motion and that the manipulations of air are secondary, or represent the material consequences of the action of Ananta Śeṣa. Similarly, one could view the phenomena described by gravitational theories as being consequences of these subtle manipulations of air.

Finally, there is the question of how a theory of gravitation should deal with the matter composing the invisible realms of the universe, including Bhū-maṇḍala. Here we confront our almost total lack of knowledge of the physics of higher-dimensional material domains.

6.C. SPACE TRAVEL

In recent years the public has received many reports of flights through outer space made by manned and unmanned vehicles launched from the United States and Russia. These include manned and unmanned orbital flights around the earth and journeys by robot vehicles to Venus, Mars, and other planets. And, of course, the most spectacular of these adventures in outer space were the Apollo flights to the moon. In this subsection we will discuss what Śrīla Prabhupāda had to say about these flights. We will begin by discussing the Vedic idea of space travel.

The Vedic literature contains many references to the idea of traveling from planet to planet through outer space. For some beings, such as great yogis and demigods, it is possible to travel from one part of the universe to another by the direct use of mystic *siddhis*. No machines are required, and the empowered being is able to transcend the constraints of ordinary space and time. However, as we mentioned in Chapter 5, machines are also used for interplanetary travel, and it would seem that many beings who are capable of traveling through space on their own also customarily make use of such machines.

We can gather from various references that these machines, which are typically called *vimānas*, or airplanes, fall into a number of different categories, including literal space ships (*ka-pota-vāyu*) and also mind ships (*ākāśa-patana*) (SB 4.12.27p). In SB 4.6.27p, *vimānas* run by mantric hymns are mentioned, and in CC AL 5.22p, it is stated that the airplanes in Satyaloka are controlled not by gross mechanical means (yantra) but by psychic action (mantra). The higher-dimensional milieu of the upper planetary systems is the natural domain of flying machines of this type. It is interesting to note that Brahmā's swan carrier is apparently a subtle mechanism of this kind, and not a sentient living entity (SB 3.24.20p). Also, even though yogis are capable of traveling through space under their own power, we read that at the time of annihilation, the yogis living on Maharloka use airplanes to escape from the fire emanating from Ananta and fly to Satyaloka (SB 2.2.26).

In SB 2.2.23p, Śrīla Prabhupāda states that it is not possible to go beyond Svargaloka or Janaloka by either gross or subtle mechanical means. This suggests that in the heavenly planets below the level of Tapoloka and Satyaloka there are classes of subtle machines that are not capable of reaching these higher realms. There are many references to the *vimānas* of the demigods, which typically seem to be used as celestial pleasure craft. These vehicles, like the demigods themselves, must operate at a level beyond the limits of our ordinary, gross senses. However, the existence of still more powerful vehicles is indicated by the story of Kardama Muni's flying city (SB 3.33.15p), and also, of course, by the accounts of transcendental *vimānas*, such as the one that carried Dhruva

Mahārāja to Vaikuṇṭha (SB 4.12.34p).

Up to the time of Mahārāja Parīkṣit, *vimānas* of the demigods regularly visited the earth (SB 1.19.18p), and persons such as Śālva would occasionally acquire remarkable flying machines by performing penances to satisfy demigods. It would seem that during this period, even materialistically inclined people were well aware of the existence of higher beings, and thus instead of trying to develop their own technology, such people would naturally turn to the demigods to satisfy their material desires. However, with the advent of the Kali-yuga, the earth (or at least the portion of the earth known to us) was placed under celestial quarantine, and access to higher planets was largely cut off (SB 2.6.29p).

It does seem, however, that flight to other planets is sometimes possible for human beings during the Kali-yuga. In SB 2.7.37 we read, "When the atheists, after being well versed in the Vedic scientific knowledge, annihilate inhabitants of different planets, flying unseen in the sky on well-built rockets prepared by the great scientist Maya, the Lord will bewilder their minds by dressing Himself attractively as Buddha and will preach on subreligious principles." According to Śrīla Jīva Gosvāmī, this remarkable verse refers to a different Kali-yuga than the present one. We gather from the nature of the rockets and the name of their designer that in this age, atheistic people of this earth had mastered some techniques of higher-dimensional travel and were able to challenge the authority of the demigods.

6.c.1. The Moon Flight

Śrīla Prabhupāda has often said that the astronauts have never actually visited the moon. Since this is a very controversial topic, we will discuss his various statements on this issue at some length. As we will see, these statements mainly fall into two categories. These are (1) that the demigods will not allow human beings to enter higher planets because human beings are not qualified to do so, and (2) that the astronauts have not experienced the celestial opulences actually existing on the moon, and therefore they could not have gone there.

In SB 1.5.18p Śrīla Prabhupāda states, "Some are trying to reach the moon or other planets by some mechanical arrangement. . . . But it is not to happen. By the law of the Supreme, different places are meant for different grades of living beings according to the work they have performed." He has said that the moon, Venus, and the sun are inaccessible to the "inexperienced scientists" because they are higher planets that can be attained only by works done in the mode of goodness (SB 2.8.14p). He has described the attempt of the scientists of this earth to reach the moon as being as demonic as the attack of Rāhu (SB 5.24.3p), and has said that such travel will be barred by Indra, who has a standard policy of preventing unqualified people from reaching the heavenly planets (SB 8.11.5p). Thus the immigration policy of the demigods is one important reason Śrīla Prabhupāda gives for why the astronauts could not have gone to the moon.

Śrīla Prabhupāda frequently uses the fact that the astronauts did not experience the celestial conditions on the moon as evidence that they did not go there. Thus he points out that the astronauts did not meet anyone on the moon, "what to speak of meeting the moon's predominating deity" (SB 4.22.54p). In SB 6.4.6p and 8.5.34p he comments that since the moon-god is the presiding deity of vegetation, there must be vegetation on the moon, and yet the scientists say that it is a barren desert. In SB 2.3.11p, 8.2.14p, and 8.22.32p, he cites the scientists' lack of knowledge of the variety of life on other planets as evidence that the moon trip failed. And in SB 10.3.27p he argues that those who reach the moon attain a life of 10,000 years, and thus the astronauts could not have gone.

Śrīla Prabhupāda makes several statements suggesting that higher-dimensional travel is needed to reach the moon. Thus in SB 1.9.45p he refers to the futility of trying to use mechanical spacecraft, and says that finer methods are needed. In SB 3.32.3p he points out that "it is not possible to reach the moon by any material vehicle like a sputnik," even though it hardly seems impossible to hurl a gross material object over a few thousand miles of space, or even several million. Finally, he indicates that to reach the orbit of the moon, it is first necessary to cross the Mānasa Lake and Sumeru Mountain (LB, p. 48). As we have already pointed out, no ordinary trajectory to the moon will pass by these particular landmarks.

We therefore suggest that when Śrīla Prabhupāda says that the astronauts did not go to the moon, he is referring to higher-dimensional travel to the celestial realm of the moon. From the Vedic point of view it is natural to interpret "travel to the moon" as travel in this sense. After all, if the moon is actually a celestial planet, then a journey to a place full of nothing but dust and rocks certainly couldn't count as a trip to the moon.

In an interview with a reporter in 1968 Śrīla Prabhupāda stressed that the human body is not suited to live in the atmosphere of the moon. When asked whether spacesuits could make up for this deficiency, he said that if we could use scientific methods to change the nature of our bodies, then we might be able to visit the moon. But he regarded this possibility as very remote, and said that the spacesuits would not be sufficient.

When the reporter asked whether the inhabitants of the moon would be visible or invisible, Śrīla Prabhupāda said that they would be "almost invisible," with subtle material bodies (CN, p. 179). This implies that the world of the demigods, including their dwellings, food, conveyances, and so on, would also be invisible to us. By definition, such a world is higher-dimensional: it is invisible to us but not to the beings living in it. To enter into it, we would indeed require more than a spacesuit: we would also need an "invisible" bodily form that could interact with the world of the lunar demigods.

This leaves open the question of whether or not the astronauts traveled in three-dimensional space to the moon that we directly perceive in the sky. We have pointed out that a higher-dimensional location can have a three-dimensional projection, just as a three-dimensional office address in New York City (given by avenue, street, and floor) has a two-dimensional projection (namely avenue and street). Thus the astronauts may have gone to the three-

Figure 18. An official NASA photograph showing the Lunar Module on the surface of the moon. Should we expect to see some signs of disturbance on the soft lunar surface caused by the exhaust of the Lunar Module's descent engine?

dimensional location of the moon without making the higher-dimensional journey needed to actually reach the kingdom of Candra. This would be comparable to visiting Vṛndāvana on the earth without being able to perceive the spiritual world that is actually there.

This is a definite possibility, although we do not know for certain whether it is true. A second possibility is that the astronauts may have been deluded by the demigods at some stage of their journey and may never have reached the gross moon planet. Thus, Śrīla Prabhupāda has suggested that the astronauts may have been diverted to the planet Rāhu (SB 4.29.69p). A third possibility, of course, is that the true story of the moon trip has been obscured by manmade illusions. Śrīla Prabhupāda has expressed doubt as to the honesty of the moon explorers, both in the *Bhāgavatam* 5.17.4p and in private conversations.

This brings us to the question of whether or not there was a moon hoax. Obviously, this is a very touchy question, and we have no definite evidence that settles it one way or another. Here we will simply give one piece of evidence suggesting that published reports of the moon landings may not have been fully honest. Figure 18 shows an official published picture of the Apollo lunar module on the surface of the moon (MSF, p. 397). The clearly visible footprints confirm the astronauts' statements that the lunar surface was soft and dusty. The rocket engine of the lunar module can be seen beneath the craft, a few feet above the surface.

As the lunar module descended to the surface of the moon, these rockets would have been firing continuously to break the vehicle's downward motion and also support its weight. Under the lunar gravitational pull (which is ⅙ as strong as the earth's gravity) the module would have weighed some 1,300 kg after expending most of its descent-stage propellant (MSF, p. 298). The question is, With the engine firing with enough power to support this much weight and break the module's fall, why do we see no disturbance caused by the rocket exhaust in the soil beneath the engine? The engine was supposedly shut down when the vehicle was about 1.52 meters above the surface (MSF, p. 300). One would think that its exhaust would have left some recognizable streaks or markings on the soft lunar soil. Yet none can be seen in this picture or in other, similar ones.

In summary, Śrīla Prabhupāda rejected the idea that men had visited the moon on the grounds that these men were not qualified to enter a higher planet and that their descriptions of their journey indicated they had not done so. He also indicated that their gross mechanical methods were not suitable for entering a higher planet. Apart from these firm conclusions, Śrīla Prabhupāda mentioned a few tentative possibilities as to what might have actually transpired on the moon flight, and he expressed general doubts as to the honesty of the people involved with space exploration. In this area there are many opportunities for cheating, and there is evidence suggesting that some cheating has taken place. However, to obtain conclusive proof of large-scale cheating would be very difficult, and possibly dangerous.

6.D. THE UNIVERSAL GLOBE AND BEYOND

According to the *Bhāgavatam*, this universe consists of a spherical inner portion four billion miles in diameter, surrounded by a series of seven coverings. In this subsection we will describe the nature and dimensions of these coverings and compare this aspect of Vedic cosmology with the modern conception of the distant regions of the universe.

Modern Western cosmologists have generally regarded the universe as having the same basic nature in all locations. One uniform geometrical framework is used to describe all space. Matter is regarded as existing in space, and it is assumed that the physical laws of our earthly laboratory experience govern the interactions between material elements in all parts of the universe. Thus the different conditions prevailing in different locations are attributed solely to the different arrangements of matter temporarily existing at those locations.

Traditionally, the geometrical framework has been three-dimensional Euclidian geometry, and thus the universe has been assumed to extend uniformly to infinity in all directions. In recent years, however, Einstein introduced four-dimensional non-Euclidian geometries, in which space can curve back on itself in a manner analogous to the curved surface of a sphere. This allowed people to formulate models of the universe in which the total volume of space is finite but there are no boundaries, and in which conditions are still essentially the same everywhere.

In Vedic cosmology the material world is not assumed to be of the same nature in all places, and space is not postulated as an absolute background within which all phenomena take place. Rather, material space, or ether, is generated at a certain phase in the process of creation, and this takes place only in certain bounded domains, called *brahmāṇḍas*. Śrīla Prabhupāda has spoken of these domains as universes and thus given a new meaning to this English word.

As we have described in Chapter 2, the Vedic literature takes the Supreme Personality of Godhead to be the ultimate source of all manifestations, and it maintains that the universes are generated by the transformation of the Lord's external energy. In the process of creation, the material elements are generated in the following order: *mahat-tattva*, false ego, mind, intelligence, sound, ether, touch, air, form, fire, taste, water, odor, and earth (SB 3.26.23–44).

Here the term *mahat-tattva* refers to the manifest form of Kṛṣṇa's total material energy, which is produced from *pradhāna*, the unmanifest or undifferentiated form of that energy (SB 3.26.10 and 17–20). The *mahat-tattva* is the source of the false ego, a material energy that serves to cover the true self-awareness of the conditioned living beings. The false ego operates in three modes, called goodness, passion, and ignorance, and thereby generates mind, intelligence, and subtle sound. Here, sound (*śabda-tanmātra*) refers not to a vibration within gross matter but to a subtle energy that generates the gross material elements and vibrates within the element of false ego in ignorance. Ether, the first element produced from this energy, is the source of the subsequent elements in our list.

When the Vedic ether is mentioned, the objection will often be raised that the idea of an ether was banished from physics by Einstein's theory of relativity. This objection refers to the classical "luminiferous ether," which was shown by the Michelson-Morley experiment to be stationary with respect to the earth (see Section 6.a). This conception of the ether was indeed rejected by Einstein, but he simply replaced it with another conception. In fact, Einstein said, "According to the general theory of relativity, space without ether is unthinkable; for in such space there would not only be no propagation of light, but also no possibility of existence for standards of space and time" (CH, pp. 53–54).

According to the Third Canto of *Śrīmad-Bhāgavatam*, ether is the basic fabric of material space. Since air, fire, water, and earth are produced from ether, these gross material elements can be regarded as transformations of space. It is interesting to note that such ideas have been recently contemplated by modern physicists. For example, the theory of geometrodynamics created by the physicist John Wheeler is an attempt to define all matter in terms of perturbations in the fabric of space. Also, the scientists working on quantum mechanical versions of general relativity are all trying, in effect, to show how the fabric of space can be derived from some kind of wave motion (or quantum wave function). This can be compared with the Vedic idea that ether is generated from subtle sound.

It is also interesting to note that in the Vedic process of creation, the sequential unfolding of the elements from ether involves an alternation of gross material substances and modes of sense perception (*tanmātras*). Thus, according to the Vedic conception, the properties of matter are intimately tied together with the processes of sense perception occurring in conscious living entities. This aspect of matter is completely disregarded in modern physics, although there is some recognition by quantum theorists such as Eugene Wigner that a complete theory of matter must take into account the existence of a conscious observer (WG).

Since our theme in this book is the structure of the universe, we will not discuss the process of creation of the elements in more detail. For us the key feature of this process is as follows: In the first step, "a part of the material nature, after being initiated by the Lord, is known as the *mahat-tattva*" (SB

TABLE 13
The Coverings of the Universe

	(1)	(2)	(3)	(4)	(5)	(6)	(7)
1			earth		earth	earth	earth
2	water		water	water	water	water	water
3	fire	fire, effulgence	fire	fire	fire	fire	fire
4	air	air	air	air		air	air
5	sky	ether	ether	sky		ether	sky
6	ego, noumenon			ego		mind	material energy
7	material nature			mahat-tattva		ego	false ego

Here we compare seven different lists of the coverings of the universe given in the *Bhāgavatam*. These are taken from: (1) SB 2.1.25p, (2) SB 2.2.28p, (3) SB 3.11.41p, (4) SB 3.26.52p, (5) SB 3.29.43p, (6) SB 3.32.9, and (7) SB 6.16.37p. (In two cases, air is listed before fire in one place and also listed in the standard order on the same page. We have taken these to be typographical errors and assumed that the standard order is correct.)

2.2.28p). The generation of false ego occurs within a restricted part of the *mahat-tattva*. Within part of this region, subtle sound becomes manifest, and then ether becomes manifest within part of the region of subtle sound. In general, each successive element becomes manifest within a small portion of the region in which the preceding element is present. This is described by Śrīla Śrīdhara Svāmī, who is cited by Śrīla Prabhupāda in this connection in SB 2.2.28p.

The result is that the material energy becomes filled with innumerable spherical regions of *mahat-tattva* and false ego. Each of these regions constitutes a particular universe, or *brahmāṇḍa*, and contains concentric spherical regions in which the successive material elements are manifest. Within the center of each of these systems of concentric globes is a hollow region containing the inhabited planetary systems of that universe.

The part of the universe in which one element is manifest but the subsequent element is not is called the universal shell or covering corresponding to that element. Generally, it is said that the inner, hollow portion of the universe is covered by seven successive shells, each ten times as thick as the one within it (SB 3.11.41). In different parts of the *Bhāgavatam* Śrīla Prabhupāda gives a number of partial lists of these different coverings. Since doubt is sometimes expressed as to what elements the various coverings consist of, we have collected together some of these lists in Table 13.

One question that is sometimes raised is, Does the first covering of the universe consist of earth or water? From this table we conclude that Śrīla Prabhupāda was generally alluding only briefly to the coverings and not trying to give an exhaustive enumeration of them. We therefore suggest that the innermost layer of the universe must be of earth, since it is listed as earth four times. In the cases where water is listed as the first covering, it may be that the earth-covering is being amalgamated with the inner, earthly region of the universe. In general, it would seem that in some lists certain layers are amalgamated together, while in others they are subdivided.

In SB 5.21.11p Śrīla Prabhupāda indicates that the coverings of the universe make it impossible for us to see the suns of other universes. We note that this should be impossible even if the layers of earth, water, and air were perfectly transparent. The reason for this is that light as we experience it is a manifestation of the fire element, and thus where there is light there is fire (SB 3.26.38–40). Therefore, it should not be possible for light from the interior of a given universe to pass beyond that universe's shell of fire. (There is light in the region beyond the universal coverings, but this is not material light, and it cannot be seen unless one has attained a certain level of spiritual advancement. Thus, the light of the all-pervading *brahmajyoti* is all around us, but it cannot be seen with ordinary vision.)

In the *Bṛhad-bhāgavātamṛta* the coverings are listed as being made of earth, water, light, air, ether, ego, and *mahat-tattva* (BB, pp. 134–35). There it is stated that variegated activities take place within each shell. Each shell is presided over by a demigoddess, beginning with the earth goddess, Bhūmi, in the first shell and ending with Prakṛti, the personified material energy, in the last. A yogi who is trying to attain liberation by leaving the material universe is presented with temptations within each shell, which he must overcome in order to continue his journey.

In SB 3.11.41p it is stated that the earthly covering of the universe is ten times the thickness of the universe itself, or 40 billion miles. This is confirmed in other places in the *Bhāgavatam*, including SB 3.29.43p. However, it is stated in SB 2.2.28p that the first covering extends "eighty million miles." This can be reconciled with the other statements about the first layer if it is a misprint and should read "eighty billion miles." In that case the figure of 80 would refer to the total thickness of the first shell along a diameter of the universe, whereas

40 billion refers to its thickness along a radius.

If we assume that the first shell has a radial thickness of forty billion miles, and that each successive shell is ten times as thick as the one preceding it, then the outer radius of the seventh shell comes to 44,444,442 billion miles. The inner region containing the planetary systems is therefore extremely small compared to the thickness of the outer coverings of the universe. According to CC AL 5.22p the universes are themselves innumerable, and they float in foamlike clusters within the unlimited Causal Ocean. Thus we can see that the idea of vast cosmic distances is present in the Vedic literature, and is not solely a product of recent cosmological thinking.

6.d.1. The Scale of Cosmic Distances

At this point the objection may be raised that although the scale of the clustered universal globes may be very large, the inner globe of this particular universe is described as being far too small to accommodate everything we can observe in the sky. It is not possible to fit even the solar system within a 2-billion-mile radius, what to speak of stars and distant galaxies. Thus, if what we can see must indeed lie within the earthly, or even the fiery, shell of this universe, then the Vedic account is seriously contradicted by modern observations.

In response to this objection we can offer the following tentative observations. In Section 4.c we observed that the rate of passage of time is much slower on Satyaloka than it is on the earth. We suggested that there might also be a comparable transformation of space in the region of Satyaloka. Thus, while a yogi traveling to Satyaloka may experience that he is crossing 2 billion miles, from our point of view he might be covering a much greater distance. We therefore suggest that when the Vedic literature speaks of a distance of 2 billion miles to the shell of the universe, it is referring to this distance as it would be perceived by the demigods, yogis, and *ṛṣis* who can actually make this trip.

In Chapter 1 we discussed a purport from *Caitanya-caritāmṛta* that is consistent with this idea. CC ML 21.84 states that the diameter of this universe is 4 billion miles. This yields a circumference of approximately 12.566 billion miles. Yet in the purport Śrīla Prabhupāda cites information from Śrīla Bhaktisiddhānta Sarasvatī indicating that the circumference of the universe is 18,712,069,200,000,000 × 8, or 149,696,553.6 billion miles. If we are to take this figure seriously, then we must accept that there exist different scales of distance that can be applied to the universe. In Chapter 1 we calculated on the basis of this figure (plus some considerations involving the length of the *yojana*) that the radius of the universe must be about 5,077 light-years. This would mean that the diameter of the fiery shell (marking the ultimate limit for the travel of material light) must be 4,442 × 5,077, or some 22.5 million, light-years, a respectable distance even by modern cosmological standards.

In SB 3.26.52p Śrīla Prabhupāda states, "The space within the hollow of the universe cannot be measured by any human scientist or anyone else." This also suggests that something unexpected must happen to space (as well as time) as one approaches the universal shell, for it hardly seems impossible to measure a distance of 2 billion miles in ordinary space. That such a transformation of space and time should occur is in agreement with the basic character of the universal coverings themselves. As one passes from covering to covering, the nature of the material manifestation is progressively transformed, until finally one emerges into a purely spiritual realm (SB 2.2.28p). Thus, it would not be surprising if transformations of the material energy and its laws of operation were to occur as one approached the first universal shell.

6.E. THE NATURE OF STARS

In modern astronomy stars are regarded as suns that are so far away from us that they appear as the minute points of light we see at night. Some stars are regarded as being as large and bright as our sun, and some are regarded as being much brighter or much dimmer. Modern astronomers have worked out an elaborate theory of the inner workings of stars, and they claim to be able to explain in detail their origin, life history, and final demise.

In contrast, Śrīla Prabhupāda has repeatedly compared the stars to reflecting planets or moons. His reasoning is presented in the purport to the verse in *Bhagavad-gītā*, where Kṛṣṇa states, "Among the stars I am the moon" (BG 10.21). There Śrīla Prabhupāda says, "It appears from this verse that the moon is one of the stars; therefore the stars that twinkle in the sky also reflect the light of the sun. The theory that there are many suns within the universe is not accepted by Vedic literature. The sun is one, and as by the reflection of the sun the moon illuminates, so also do the stars. Since the *Bhagavad-gītā* indicates herein that the moon is one of the stars, the twinkling stars are not suns but are similar to the moon."

In BG 15.12 it is directly said that the sun illuminates the entire universe, and Śrīla Prabhupāda comments, "From this verse we can understand that the sun is illuminating the whole solar system. There are different universes and solar systems, and there are different suns, moons, and planets also, but in each universe there is only one sun." A similar statement is made in BG 13.34, and Śrīla Prabhupāda speaks of the unique position of the sun and the moonlike nature of the stars in SB 3.15.2p, 4.29.42–44p, and 5.16.1p, as well as in TQK, p. 102.

It is clear that from the viewpoint of demigods and yogis, all the stars and planets of the universe lie within a fairly small neighborhood and can be

TABLE 14
The Lunar Mansions

Part of Śiśumāra-cakra	Nakṣatra	Western Star Name
n	Revatī	Zeta Piscium
n	Aśvinī	Alpha Arietis
n	Bharanī	Musca
n	Kṛttikā	Pi Tauri, Pleiades
n	Rohiṇī	Alpha Tauri, Aldebaran
n	Mṛgaśīrṣa	Lambda Orionis
right foot	Ārdrā	Alpha Orionis
loins	Punarvasu	Beta Geminorum
loins	Puṣya	Delta Cancri
left foot	Āśleṣā	Alpha 1 & 2 Cancri
s	Maghā	Alpha Leonis, Regulus
s	Pūrva-phalgunī	Delta Leonis
s	Uttarā-phalgunī	Beta Leonis
s	Hasta	Gamma or Delta Corvi
s	Citra	Alpha Virginis, Spica
s	Svāti	Alpha Bootis, Arcturus
s	Viśākhā	Alpha or Xi Libra
s	Anurādhā	Delta Scorpionis
left shoulder	Jyeśṭhā	Alpha Scorpionis, Antares
left ear	Mūla	Nu Scorpionis
left eye	Pūrvāṣādhā	Delta Sagittarii
left nostril	Uttarāṣādhā	Tau Sagittarii
right nostril	Abhijit	Alpha Lyri
right eye	Śravaṇā	Alpha Aquilae
right ear	Dhaniṣṭhā	Alpha Delphini
right shoulder	Satabhiṣā	Lambda Aquarii
n	Pūrvabhādra	Alpha Pegasi
n	Uttarabhādra	Alpha Andromedo

The central column lists the 28 *nakṣatras*, or lunar mansions. The column on the right lists the Western names for their principal stars. On the left are the parts of the body of the *śiśumāra-cakra* represented by these stars. The *n*'s represent the right side and the course of the sun to the north; the *s*'s represent the left side and the course of the sun to the south.

reached by interplanetary travel. Thus, the stars in the Kṛttikā constellation (corresponding to the Pleiades) are associated with the wives of the moon-god (SB 6.6.23), and the seven stars of the big dipper are associated with the seven sages. (We also read in SB 1.9.8p that Candramāsī, the wife of Bṛhaspati, was "one of the reputed stars.")

In SB 5.22.11 it is stated that 28 important stars headed by Abhijit are located 200,000 *yojanas* above the moon. This distance seems short indeed, but we should consider that in this verse the word *nakṣatra*, or star, has a special meaning. In Vedic astronomy there are 28 important constellations, headed by Abhijit. Of these, 27 lie along the ecliptic and are used to divide it into 27 equal units of 13⅓°. These constellations are referred to as *nakṣatras*, or lunar mansions. They are particularly connected with the motion of the moon, since the moon completes one orbit in about 27.3 days. In SB 5.22.5 the *nakṣatras* are referred to in the following statement: "According to stellar calculations, a month equals two and one quarter constellations." (Note that 2¼ times 13⅓° equals 30°.)

The 28 *nakṣatras* are mentioned in the description of the *śiśumāra-cakra* in Chapter 23 of the Fifth Canto. The *śiśumāra-cakra* is an imaginary form in the heavens that is made up of constellations and visualized as a gigantic animal. This form is worshiped by some yogis as a manifestation of the *virāṭ-rūpa*, or the external form of Kṛṣṇa. Table 14 lists the 28 *nakṣatras* and the Western (Greek and Arabic) names for their principal stars, or *yoga-tāras*. These identifications are from SS, p. 62. We have also indicated the different parts of the *śiśumāra-cakra* that these *nakṣatras* represent. These are taken from SB 5.23.7.

Apart from the 28 *nakṣatras*, the only stars for which distances are given in the *Bhāgavatam* are the planets of the seven sages, which are said to lie 1,100,000 *yojanas* above Saturn, and the polestar, Dhruvaloka, which is said to be 1,300,000 *yojanas* above these planets (SB 5.22.17 and 5.23.1).

These distances, of course, are also very small (and as we have indicated in Chapter 5, they should be interpreted as heights perpendicular to the plane of Bhū-maṇḍala). They conform to the idea that the stars in general are fairly close by, from the point of view of the demigods, that they are planets reflecting the light of the sun, and that the sun has the unique role of illuminating the entire universe.

This does not mean, however, that the distances to the stars as they appear to us will necessarily be this small. The distances may seem larger to us than they would to a demigod who was actually traversing them. As we have already indicated, the higher modes of travel used by the demigods may involve transformations of both space and time that make the distances shorter for them than they would be for a manmade machine traveling in the ordinary three-dimensional fashion. Thus, it might be that a spaceship launched from the earth toward the polestar would actually have to travel for many years at nearly the speed of light to get there.

In SB 3.15.26p Śrīla Prabhupāda makes an interesting remark: "By present standards, scientists calculate that if one could travel at the speed of light, it would take forty thousand years to reach the highest planet of this material world. But the yoga system can carry one without limitation or difficulty." If the distances to the stars are really very short, one might ask why Śrīla Prabhupāda would apparently give credence to this example of the modern idea of interstellar travel. It makes perfect sense to do so, however, if the distances as experienced by a three-dimensional traveler are very large, whereas the distances experienced by a yogi are relatively small.

At this point one might object that if the ordinary, three-dimensional distances to the stars are very large, then the inverse square law for the diminution of light intensity with distance implies that the stars must be shining very brightly. For the stars to appear as bright as they do to us, they must actually be shining as brilliantly as suns. Furthermore, the fact that the light of the stars has an emission spectrum shows that they are actively generating light and not just passively reflecting it.

In response to this objection, two points should be made. The first is that it is not necessary to suppose that stars do not generate their own light. Śrīla Prabhupāda compares the stars to moons, but he also gives an "educated guess" to the effect that there are mild and pleasing flames on the moon that generate illumination (SB 5.20.13p). Thus the conclusion is that stars may be fiery and thus generate an emission spectrum, but they are not independent suns. Indeed, Śrīla Prabhupāda has said, "The stars may have the same composition as the sun, but they are not suns" (letter to Svarūpa Dāmodara dāsa, Nov. 21, 1975).

The second point is that the inverse square law for the propagation of light may not hold universally. If that is the case, then we cannot conclude that if a star is at a distance of many light-years, it must therefore be as brilliant as the sun. In general, we propose that it cannot be taken for granted that the laws prevailing in remote parts of the universe are the same as the laws that hold here on the earth. The Vedic literatures describe phenomena on the higher planets that are quite different from the phenomena we experience on the earth, and they also indicate that the operation of the material energy on the earth was significantly different in earlier *yugas* (SB 1.4.17p). This suggests that laws governing the production and propagation of light might also be different in different parts of the universe. Of course, if the laws of physics are different in different parts of the universe, then it might also be that stars appear to be more distant than they actually are. It may even be that the very idea of distance as we know it breaks down in remote regions of the universe. Once we allow the laws of physics to vary, the possibilities are limitless.

In the next chapter we will show that in modern cosmology, there is abundant evidence indicating that the laws of physics may indeed change significantly as one travels from the earth to remote regions of the universe.

Red Shifts and The Expanding Universe

In this chapter we enter into the domain of extra-galactic cosmology and discuss the big bang theory and the evidence for an expanding universe. We illustrate how scientists construct theoretical models that are mistaken for reality by millions of people who encounter them in authoritative textbooks and in popular presentations. We show, however, that even within the narrowly defined domain of astrophysics, new observations may arise that fundamentally challenge these models and reveal them to be mere systems of speculative ideas. In the example that we consider, this fundamental challenge shows the need for drastic revisions in the modern understanding of the universe as a whole—revisions that thus far have not been seriously considered in the scientific community.

Let us begin with the theory of the big bang. Basically, this is the idea that in the beginning (or before the beginning, if you will), all matter in the universe was concentrated into an infinitely small volume at an infinitely high temperature and pressure. Then, according to the story, it exploded with tremendous force. From this explosion rushed a superheated, ionized gas, or plasma. This plasma expanded uniformly until it cooled sufficiently to form ordinary gas. Within this cooling cloud of expanding gas formed galaxies, and within the galaxies took birth generations of stars. In turn planets such as our own earth formed around the stars.

But here's a fact that few people realize: Even with the most powerful telescopes, it is not possible to actually see galaxies moving away from us. The images we see are static, and scientists would not expect them to show visible motion, even if observations could go on for centuries.

So how do we really know the universe is expanding? All we have to go on is the light and other kinds of radiation that travel to us from across the reaches of interstellar space. Images formed from this radiation do not directly show universal expansion, but subtle features of the radiation have convinced scientists that this expansion is taking place. What scientists do is first assume that the earthly laws of physics apply without change throughout the universe. They then try to figure out how processes obeying these laws could produce the observed light.

To understand how scientists have used this way of analyzing light to conclude that the universe is expanding, let us go back into the history of astronomy and astrophysics. Examining the heavens, astronomers long ago observed that in addition to individual stars and planets, there were many faintly glowing bodies in the sky. They called them nebulae, a Latin word meaning "clouds," and later on, as their conceptions evolved, they called them galaxies.

Larger than the full moon in the night sky, yet so dim that it is hardly visible to the unaided eye, is the nearby galaxy Andromeda. In the early part of this century, astronomers turned powerful new telescopes on this and other galaxies and found that they appeared to be vast islands of billions of stars. At further distances are found entire clusters of galaxies.

Until the discovery of stars in Andromeda, it was generally thought that all celestial bodies were located within the boundaries of our local Milky Way galaxy. But with this development and the discovery of other, more distant galaxies, all that was changed. The dimensions of the universe expanded beyond comprehension.

Up until the early part of this century, scientists believed that the basic objects in the universe were static in relation to one another. Then in 1913 the American astronomer Vesto Melvin Slipher came to study the spectra of light coming from a dozen prominent nebulae and concluded that they were moving away from the earth at speeds of up two million miles per hour.

How did Slipher reach this astonishing conclusion? For some time, astronomers had been using spectrographic analysis to determine the elements present in the stars. It was known that the spectrum of light associated with a particular element will show a characteristic pattern of lines that serves as a kind of signature for the element.

Slipher noticed that in the spectra of galaxies he studied, the lines for certain elements were shifted toward the red part of the spectrum. This curious phenomenon is called a "red shift." Slipher interpreted the red shift as a Doppler effect, indicating that the galaxies were moving away. This was the first major step toward the idea that the entire universe is expanding. (If the lines in the spectrum had been shifted toward the blue end of the spectrum, that would have indicated that the galaxies were moving toward the observer.)

The Doppler effect is often explained by using the example of a train whistle, which seems to change pitch as the train goes by. This phenomenon was first scientifically studied in 1842 by Christian Johann Doppler, an Austrian physicist. He proposed that the intervals between the sound waves emitted from an object moving toward a listener are compressed, causing the sound to rise in pitch. Similarly, the intervals between sound waves reaching a listener from a source moving away are elongated, and thus the sound's pitch is lower. It is reported that Doppler tested this idea by placing trumpet players on a flatcar drawn by a locomotive. Musicians with perfect pitch listened carefully as the trumpet players moved by them, and they confirmed Doppler's analysis.

Doppler predicted a similar effect for light waves. For light, an increase in wavelength corresponds to a shift toward the red end of the spectrum. Therefore the spectrum of an object moving away from an observer would tend to be shifted toward the red. Slipher chose to interpret his observations of galaxies in this way, as a Doppler effect. He noted a red shift and decided the galaxies must be moving away.

Another step toward belief in an expanding universe took place in 1917, when Einstein published his theory of general relativity. Before Einstein, scientists had always assumed that space extended to infinity in all directions and that the geometry of space was Euclidean and three-dimensional. But Einstein proposed that space could have a different kind of geometry-four-dimensional curved space-time, in which space could curve back on itself.

There are many forms that space could take, according to Einstein's theory. One is a closed space without a boundary, like the surface of a sphere; another is a negatively curved space that extends to infinity in all directions.

Einstein himself thought the universe should be static, and he adjusted his equations to insure this outcome. But almost immediately, Willem de Sitter, a Dutch astronomer, found solutions to Einstein's equations that predicted a rapidly expanding universe. The geometry of space would change with time.

7.A. HUBBLE'S EXPANDING UNIVERSE MODEL

De Sitter's work caused a stir among astronomers around the world. One of them was Edwin Hubble. Hubble had been present when Slipher had announced his original findings about the motion of galaxies to a meeting of the American Astronomical Society in 1914. In 1928 Hubble set to work at the famous Mt. Wilson observatory in an effort to bring together De Sitter's theory of an expanding universe and Slipher's observations of receding galaxies. Hubble reasoned like this: In an expanding universe you would expect the galaxies to be moving apart from each other. And the further apart from each other they were, the faster they should be moving apart. This would mean that from any point, including the earth, an observer should see that all other galaxies are moving away and that, on the average, the further away a galaxy is, the faster this motion should be.

Hubble set out to see if this were true and discovered that there seemed to be a proportional relationship between the distance of galaxies and the degree of their red shifts. Most galaxies, he observed, had red shifts, and the greater the distance, the greater the red shift.

This raises a vexing question: How did Hubble know how far away any given galaxy was? That was a very difficult problem for Hubble, and it remains so even for today's astronomers. After all, there are no measuring rods that can reach to the stars. But the basic idea is this: We can begin by using various methods to estimate the distances of nearby stars. Then, proceeding step by step, we can build a "cosmic distance ladder" that gives us estimates of the distances of a few galaxies. If we can find a way of guessing the intrinsic brightness of galaxies, we can then relate unknown galactic distances to known ones by making measurements of apparent galactic brightness. This is according to the inverse square law.

Here we will not go into the details of the complex procedures used to establish this distance ladder. Suffice it to say that they involve many theoretical interpretations that are fraught with uncertainty and subject to revision, often in unexpected ways. This will emerge as we go along.

Hubble, using his methods of approximating distance, established a proportional relationship, now known as Hubble's law, between degree of red shift and distance for galaxies. He believed he had clearly shown that the galaxies most distant from us had the biggest red shifts and were thus receding from us most rapidly. This he took as ample evidence that the universe is expanding.

Eventually this idea became so solidly established that astronomers began to apply it in reverse: If distance is proportional to red shift, then one can measure the distance of galaxies simply by measuring their red shifts.

But as we have noted, Hubble's distance figures are not direct, accurate measurements of how far away galaxies are. Rather, they are derived indirectly from the apparent brightness of the galaxies. Thus the expanding universe model has two potential defects: First, the brightness and dimness of celestial bodies could quite possibly be caused by something other than how far away they are, and thus the distance figures derived from them could be flawed. And second, it is possible that the red shift might not be connected to velocity.

In fact, a number of astronomers are convinced that some red shifts are not caused by a Doppler effect. And some even go so far as to question the very concept of an expanding universe.

7.B. ANOMALOUS RED SHIFTS: THE OBSERVATIONS OF HALTON ARP

One astronomer who doubts the interpretation of all red shifts as Doppler effects is Halton Arp, who has served on the staff of the Hale Observatory at

Mt. Palomar and is currently doing research at the Max Planck Institute near Munich, West Germany. At Palomar, Arp observed many examples of discordant red shifts that do not follow Hubble's law. His analysis suggests to him that red shifts in general may be due to something other than a Doppler effect. Here we should ask why scientists generally interpret red shifts as being caused exclusively by the Doppler effect. It may be true that a Doppler effect produces a red shift, but how do we know that a red shift must be due to a Doppler effect? One of the main reasons for this conclusion is that, according to modern physics, the only phenomenon other than a Doppler effect that will produce a pronounced red shift is a powerful gravitational field. If light is going up against a gravitational field, it loses energy and undergoes a red shift. However, astronomers don't find this explanation applicable for stars and galaxies, because the fields would have to be of incredible strength to produce the observed red shift.

Arp argues that he has found objects with high red shifts in close proximity to ones with low red shifts (AR1-4, RC). According to the standard expanding-universe theory, an object with a small red shift should be relatively close to us, and an object with a large red shift should be far away. Thus, two objects that are relatively close to each other should have similar red shifts.

But Arp gives the following example: The spiral galaxy NGC 7603 is connected to a companion galaxy by a luminous bridge, yet the companion galaxy has a red shift 8,000 kilometers per second higher than that of the spiral galaxy. Judging by the disparity in their red shifts, the galaxies should be at vastly different distances—to be precise, the companion should be about 478 million light-years further away—yet strangely, the two galaxies seem close enough to be physically connected. For comparison's sake, our own galaxy, the Milky Way, is said to be just 2 million light years from its nearest neighbor, the galaxy Andromeda.

Of course, there are some defenders of the standard view who strongly disagree with Arp's interpretation. John N. Bahcall, of Princeton's Institute of Advanced Studies, maintains there is no reason to suppose that the two galaxies are connected (RC). The objects are actually distant from each other and just appear to be closely associated. The so-called luminous bridge is there, but the more distant galaxy just happens to be lined up behind it from our point of view.

To illustrate his criticism, Bahcall gives this specific rebuttal: He shows a photograph of a star within our own Milky Way galaxy apparently connected to a distant galaxy by what appears to be a luminous bridge. Are they connected? Bahcall points out that this is clearly impossible because the star is a bright foreground star in our own galaxy, while the distant galaxy is 44 million light-years away.

However, Arp responds by saying that Bahcall is just being frivolous. The galaxy he shows is not in any way unusual. The luminous bridge to the star is simply one of its normal spiral arms. But in the example Arp himself has chosen, the bridge is an unusual structure, not normally found in such galaxies. The likelihood that two galaxies of this type could be found in such a relationship is far less than the likelihood that a star in the Milky Way will be lined up with an ordinary galaxy.

Arp has found many other examples that seem to violate the traditional understanding of the red shift. Here is one of the most controversial of these discoveries: Near the spiral galaxy NGC 4319 is a quasar, Makarian 205, apparently connected to the galaxy by a luminous bridge. The galaxy has a red shift of 1,800 km per sec, giving it a distance of about 107 million light-years. The quasar has a red shift of 21,000 km per sec, which should mean that it is located 1.24 billion light-years distant. But Arp suggests that they are definitely connected and that this shows that the standard interpretation of the red shift is wrong in this case. (We may note, by the way, that the very fact that astronomers express red shifts in terms of kilometers per second shows their commitment to the idea that the red shifts are Doppler effects.)

Critics took their own photographs of NGC 4319 and claimed not to have found the connecting bridge shown in Arp's picture. Others said the bridge was a "spurious photographic effect." But recently, Jack M. Sulentic, of the University of Alabama, did extensive photometric studies of the two objects and concluded that the connecting bridge is real (SU).

Another example of discordant red shifts noted by Arp is found in the highly unusual chain of galaxies called Vorontsov-Velyaminov 172, after its Russian discoverers. In this chain, the smaller, more compact member has a red shift twice as great as the others.

In addition to pairs of galaxies with discordant red shifts, Arp points out something even stranger—it appears that quasars and galaxies can eject other quasars and galaxies. Here are some examples: The exploding galaxy NGC 520 has a fairly low red shift. Located along a straight line running to the southwest from the galaxy are 4 quasars of the faint type. Arp says that these faint quasars are the only ones in this region. Could it simply be an accident that they are arranged almost exactly on a straight line from the galaxy? Arp says the chances of this are extremely remote and suggests that the quasars were ejected from the exploding galaxy.

Interestingly enough, the quasars have much larger red shifts than the galaxy that seems to be their parent. This is remarkable, since according to the standard theory of the red shift, the quasars should be much further away than the galaxy. Arp interprets this and other, similar examples by proposing that freshly ejected quasars are born with high red shifts, which gradually decrease as time passes.

Some scientists question whether it is really possible for galaxies to eject other massive objects such as galaxies or quasars. In response, Arp points to a striking photograph of the giant galaxy M87 ejecting a jet of material. When we look at the galaxies of the elliptical type in the region around galaxy M87 (which is also elliptical), we find that they all fall on a line drawn in the direction of the jet of ejected material. This suggests to Arp that these galaxies have been ejected by M87.

How is it that a galaxy can emit another galaxy? If a galaxy is an "island universe" consisting of a vast aggregate of stars and gas, how can it emit another galaxy, which is a similar aggregate of stars and gas?

It has been argued that radioastronomy may provide a clue. In recent times, radioastronomers have agreed that vast radio-emission areas can be ejected from galaxies. These emission areas exist in pairs on either side of some galaxies. To explain this, astronomers have postulated gigantic spinning black holes in the centers of the galaxies that gobble up nearby stars and spit out material in both directions along their axis of spin. However, if Arp's analysis is correct, one has not only to explain the radiating emission regions, which may be composed of a thin gas, but also how entire galaxies or precursors of galaxies might come flying out.

Regarding the red shifts of such ejected galaxies and quasars, Arp has found the following: The ejected objects, although in close proximity to the parent objects, have much higher red shifts. Arp maintains this can only mean that their red shifts are not due to the Doppler effect. That is, they do not measure the speed at which the object is receding. Rather, the red shift has something to do with the actual physical state of the object.

But the present laws of physics provide no inkling of what this state might be. Since a galaxy is thought to be composed of many individual stars plus clouds of dust and gas, what qualities could it have that would result in a red shift not due to velocity or gravitation? This cannot be explained in terms of known physical principles.

This seems to call for new physics. But that opens up a whole Pandora's box, because modern cosmology is completely committed to the assumption that everything we see in the universe can be explained by the known laws of physics. If the physical laws are fundamentally changed, then all the models based on them are brought into question.

Of course, Arp's findings are very controversial, and many astronomers doubt that the associations between galaxies and quasars he speaks of could actually be real. But this is only one line of evidence suggesting that the standard interpretation of galactic red shifts might be in need of revision.

7.C. HUBBLE'S CONSTANT AND TIRED LIGHT

Another line of evidence involves Hubble's constant, which is the very heart of the expanding universe model. As we have seen, according to the big bang model, the further away a galaxy is, the faster it should be going. According to Hubble's law, the speed of recession should be equal to the distance multiplied by a number called Hubble's constant. With this law, it becomes possible for astronomers to calculate the distance of galaxies simply from their red shifts. Find the red shift and divide by Hubble's constant—and now you have the distance.

The constant also gives astronomers the size of the universe. They can measure the red shift of the most distant celestial object and use the Hubble constant to determine its distance. The Hubble constant is therefore an extremely crucial number. For example, if you double the constant, you double the estimated size of the universe. Clearly, a precise value for Hubble's constant is essential for determining the size of the universe with any accuracy.

Over the years, however, different scientists have obtained many different values for Hubble's constant. The constant is expressed in kilometers per second per megaparsec. (A megaparsec is a unit of cosmic distance equal to 3.3 million light-years.) In 1929 the value of Hubble's constant was 500. In 1931 it was 550. In 1936 it was 520 or 526. In 1950 it was given as 260, down significantly. By 1956 it had dropped to 176 or 180. In 1958 it fell much further down, to 75, but in 1968 it bounced back up to 98. In 1972 it ranged from 50 all the way up to 130. In 1989, the Hubble constant is pegged at 55. All this change led one wry astronomer to say that perhaps the Hubble constant should better be called the Hubble variable.

Of course, these changes over the decades can be explained by arguing that that scientists have improved their methods and refined their calculations. But even so, something appears to be amiss.

This brings us to the work of Jean Pierre Vigier, a French astrophysicist at the Institute Henri Poincare (VG1-5). Vigier points out that even today, different observers obtain different values for Hubble's constant. Tammann and Sandage give 55 plus or minus 5. Abell and Eastmond arrive at 47, plus or minus 5. Then there is van den Bergh, who calculates a value between 93 and 111. Heidmann got 100 for his figure. De Vaucouleurs came up with 100 plus or minus 10.

If the universe is expanding according to some uniform law of proportionality, how is it that so many observers obtain so many greatly different values for the rate of expansion?

Vigier notes that when astronomers take measurements in different directions, they find different rates of expansion. He then points out something even stranger: The sky can be divided into two sets of directions. The first is the set of directions in which many galaxies lie in front of more distant galaxies. The second is the set of directions in which there are distant galaxies without foreground galaxies. Call the first set "area A," and the second set "area B."

Vigier found that if you restrict yourself to the distant galaxies in area A and calculate Hubble's constant, you get one value, and in area B you get a significantly different one. This suggests that the rate of expansion varies depending on whether we measure galaxies with or without foreground galaxies. If the universe is expanding, what could these foreground galaxies possibly have to do with the rate of expansion? Vigier suggests that in fact the measured red shifts of the distant galaxies are not caused by the expansion of the universe at all. Rather, they are caused by something quite different-something called a tired-light mechanism.

According to Vigier, as light moves through space it becomes red shifted simply from traveling a certain distance. This happens in accordance with physical laws, just like any other phenomenon. There is a law requiring that as light travels, it shifts toward the red. The effect is so small that it cannot be readily measured in laboratories on earth, but as light moves the vast distances between galaxies, the effect becomes apparent.

This is called the tired-light hypothesis because the light loses energy as it moves through space. And the more tired it becomes, the redder it becomes. Red shift is therefore proportional to distance, not to the velocity of the object. Vigier pictures the universe as not expanding. All the galaxies are more or less stationary. The red shift is not a Doppler effect; it has nothing to do with the velocity of the light's source. The red shift is caused by an inherent property of the light itself, namely that it becomes tired after traveling long distances.

Most astronomers reject the idea of tired light. In the words of Joseph Silk, of the University of California at Berkeley, "Tired light cosmologies are unsatisfactory because they invoke a new law of physics" (SK).

But Vigier presents his tired-light theory in a way that does not require radically new physics. He proposes that there is a kind of particle in intergalactic space that interacts with light in such a way as to steal energy from it. In the vicinity of massive objects, there are more of these particles than elsewhere. Using this idea, Vigier explains the different red shifts for the A and B regions in the following way: The light passing through foreground galaxies encounters more of these particles and therefore loses more energy than light not passing through regions with foreground galaxies. Thus there is a greater red shift for the light going through regions with foreground galaxies, and this accounts for the different values found for the Hubble constant.

Vigier also cites additional evidence for nonvelocity red shifts. For example, if the light from stars is measured when passing near the sun, it displays a higher red shift than when measured in a different area of the sky. Such measurements can be made only during total eclipses of the sun, when stars near the solar disc become visible in the darkness.

In short, Vigier explains the red shift in terms of a nonexpanding universe in which light behaves somewhat differently than it is normally supposed to behave. Vigier claims that his model fits the astronomical data better than the standard expanding-universe model, which cannot explain the widely different values obtained for the Hubble constant. According to Vigier, nonvelocity red shifts may be a general feature of the universe. The universe could very well be static, and thus there would be no reason for the big bang theory.

7.D. QUASARS

Doubt has also been cast on the expanding universe theory by the study of quasars, or quasi-stellar radio sources. Quasars look like stars but have very big red shifts, and thus they are considered the most distant objects in the universe, more distant than the most distant galaxies. We have already seen that Halton Arp believes some quasars are cosmologically close to us, even though they have high red shifts. Arp has also noted that many quasars tend to be located in the same vicinity of the sky as a large group of galaxies relatively close to our own. This suggests to him that the quasars may be associated in some fashion with these local galaxies and thus be at the same distance.

This raises a question: If some quasars are actually close, and thus have large nonvelocity red shifts, why couldn't that be true of quasars in general? In fact it has long been observed that there are severe difficulties with the idea that quasars are at their cosmological distances, that is, that they are at the distance obtained by applying the Hubble constant to their extremely large red shifts.

The big problem is that quasars are very bright. If they are in fact extremely far away, that means that many quasars are putting out hundreds of times more energy than the brightest galaxies, which are composed of hundreds of billions of stars. If quasars were as big as galaxies, that might not be implausible. But it turns out that quasars can vary in their light intensity over periods as short as days. This observation suggests to astronomers that they are very small compared to galaxies. No one can understand how such a small object can generate so much energy, at least by presently known physical laws.

One interesting approach to the interpretation of quasars has been proposed by Y. P. Varshni, a physicist at the University of Ottawa in Canada (VR1-3). He supports Arp's contention that quasars have nonvelocity red shifts, citing as evidence certain patterns in the way these red shifts are distributed.

Normally one would expect celestial objects like quasars to have a wide variety of red shifts with no discernible pattern. But Varshni finds that these red shifts tend to fall into well-defined groups. Each red-shift group is represented by quasars distributed widely across the sky, and very few quasars have red shifts that would place them outside the major groupings. A similar phenomenon was also noted by the astronomer Geoffrey Burbidge, who observed that an unexpectedly large percentage of quasars have red shifts grouped closely around 1.95 (BR1). (The red shift of 1.95 is expressed in terms of shift in wavelength; it comes to about 238,160 km per sec, or 79 percent of the speed of light.)

This clustering of red shifts is a very difficult phenomenon to explain. Let us apply the standard cosmological interpretation to the distance of the quasars. All of the quasars with the same red shift should be at the same distance. Thus the quasars with a red shift of 1.95 should all lie close to a spherical shell with a radius corresponding to this red shift. The same should hold true of the other red shift groupings, each of which includes quasars in a wide variety of directions. This means that the quasars lie on a series of spherical shells centered on the earth.

This conclusion is unacceptable to modern cosmological thinking because it places the earth in a special central position in the universe. There is only one center in an array of concentric shells. In effect, the earth must be at the center of the universe.

The odds that this arrangement of shells could happen by chance are next to nothing, and Varshni argues that the conclusion that the earth really is at the center of concentric shells of quasars is not acceptable. Therefore the red shifts of the quasars must be due to something other than the Doppler effect, as described in the expanding-universe model. If they are not due to the Doppler effect, they do not represent distance, and if they do not represent distance, it is no longer necessary to suppose the quasars are arranged in shells.

Varshni believes that quasars generate light in an unexpected way, a way that gives the appearance of Doppler-shifted light. According to Varshni, laser effects in the quasars give light inherently different characteristics that have nothing to do with velocity. Varshni believes scientists have mistaken the spectral lines in this type of light for Doppler-shifted spectral lines in ordinary ionized gas. So according to Varshni, the quasars are close by, and the idea that they are far away results from misinterpreting their laser-generated light as Doppler-shifted ordinary light.

Varshni's theory may or may not be true, but his observation that the spectral lines of quasars fall into definite groupings does call into question the standard theory of cosmic distances—at least for quasars. If the spectral lines are taken to be displaced by Doppler shifts and one applies the standard theory, one gets the unacceptable result that the earth is the center of the universe. If this were accepted, scientists would have to return to an idea they have consistently rejected since the time of Galileo and Copernicus.

7.E. QUANTIZED RED SHIFTS

Yet Varshni's observations represent only one of a number of strange patterns that emerge when modern astronomical data are closely examined. Another interesting pattern has been discerned by William G. Tifft, an astronomer at Steward Observatory, at the University of Arizona at Tucson (TF1-7). His conclusions have perhaps the most disturbing implications of all for the expanding-universe model.

Tifft has observed that the red shifts associated with galaxies tend to be quantized. What this means is that red shifts tend not to be just any numbers but rather multiples of a certain basic unit of about 72 kilometers per second. In general, his studies show that red shifts of galaxies are grouped at 72 kps, 144 kps, 216 kps, 288 kps, and so on.

Let us consider a pair of galaxies close to each other in space. According to Newtonian gravitational theory, these galaxies should be attracting each other gravitationally. Thus they should be orbiting around each other, falling together, or flying apart, and this relative motion should be revealed by a measurable red shift.

Tifft examined the relative red shift of many pairs of galaxies. This value, according to standard theory, would represent not the speed at which the pair is receding from the earth but rather the speed at which one galaxy is moving in orbit around the other, measured along the line of sight from the earth. Simply put, the speed is calculated as follows: The observer measures the red shift of each galaxy in a pair of galaxies. For example, one galaxy may have a red shift of 7,500 kps, and the other may have one of 7,000 kps. This means that one galaxy is at that time moving relative to the other at a speed of 500 kps along the line of sight. But because this speed is due to orbital motion, it will vary according to the positions of the galaxies at different points in time. For example, when the galaxies are moving perpendicular to the line of sight, the relative motion will be zero, and they will have exactly the same red shift at that point.

So if the two galaxies are in fact moving in orbit, the relative red shift will vary smoothly within a definite range of values. Of course, it is not possible to measure this variation for a single pair of galaxies. They would not display any visible motion or change of red shift within the lifetime of the observer. Therefore it is necessary to observe hundreds of pairs of galaxies and calculate their relative red shifts. If we did this, we would expect to find a nearly continuous spread of values, because we would be catching the galaxies at a variety of orbital positions relative to our line of sight.

But Tifft has found this not to be the case. The red shifts are grouped in near multiples of a basic unit—72 kilometers per second. This indicates to Tifft that the measured red shift is a nonvelocity red shift and that the galaxies in each pair are actually not orbiting each other. One might argue that perhaps the red shifts are caused by something other than the Doppler effect, but surely the galaxies must still be orbiting one another. But Tifft points out that even if something other than relative velocity is causing red shifts, orbital motion should still produce a smooth distribution of Doppler-effect red shifts in addition to this. But this is not what he finds.

Tifft's findings apply not only to galaxies moving in pairs, but to whole groups of galaxies. This poses two questions that modern physics cannot answer. The first is, How is it possible for galaxies to have a nonvelocity red shift? Tifft proposes that it is caused by the nature of the galaxies themselves. They produce light that is red shifted because of internal properties having to do with some as-yet unknown law of nature. The second question is, If the red shift is not due to motion, then what is the motion of the galaxies? If they are orbiting, then there should be a continuous range of Doppler shifts, whatever the internal properties of the galaxies might be. Could it be that they are not orbiting? Then, according to Newton's or Einstein's laws of gravity, they should be falling together or perhaps flying apart. They should still be moving relative to one another, but the indication is that they are not. Therefore, according to Tifft, new principles of gravitation are necessary.

There is already evidence that might be interpreted as indicating that Newton's laws may have to be revised, especially in relation to galaxies. For many years, scientists have found great difficulty in accounting for the dynamics of galactic motion in terms of the law of gravity. For example, it may be seen that certain galaxies appear to be orbiting in a cluster, but the dynamics of mass and gravity suggest they should not be arranged in that way. Given their supposed velocities, they would have to be much more massive in order to orbit. But rather than sacrifice the laws of gravity, astronomers have posited the existence of great quantities of invisible dark matter to account for the missing mass. Some say 90% of the mass of the universe is missing.

But another way to deal with this issue is to say that the laws of gravity need revision, and Tifft is suggesting this, based on his research. With new laws of gravity, the need to posit missing mass disappears. Is Tifft right or wrong? As of now, it isn't possible to say. But his ideas do show how scientists, operating with the very limited data they have been able to acquire, are running into all kinds of contradictions in their attempt to comprehend the universe.

Thus far we have discussed pairs and groups of galaxies. We have seen how their red shifts, representing movement relative to one another along an observer's line of sight, should vary smoothly through a wide range of values. But Tifft has found that the red-shift values are quantized in multiples of a constant unit, and thus he concludes that they are not moving at all relative to one another.

But what about the galaxies' absolute movement along our line of sight? Is it possible that the galaxies are also standing still in relation to us—that they are not moving away, as the expanding-universe model tells us they should be?

Tifft argues that they are not moving. If they are moving due to expansion of the universe, their red shifts should span a wide range, covering all possible intermediate values. But Tifft proposes that these red shifts are also quantized. Normal measurements do not show this, but Tifft points out that when the effect of solar motion is subtracted, the quantization of the red shifts becomes unmistakably clear. The red shifts do not vary smoothly but instead come in multiples of a constant number.

Let's take a closer look at this problem. If the red shifts are quantized, as Tifft says they are, then the sun, because of its motion, adds a Doppler effect to those quantized red shifts. What is added will depend on the angle of the distant galaxy's motion relative to the sun's motion. If the galaxy is moving perpendicular to the sun's path, the sun's movement will not add anything. At 0° there would be a negative red shift (i.e., a blue shift), which would be subtracted. At 180° one would add a positive red shift. At points in between, one gets other values. And by adding these values, one breaks up the quantized nature of the red shifts.

To detect the quantization, one has to subtract the red shift due to the sun's motion from the observed red-shift values. Tifft says he has done just that, and has found that galaxies have red shifts arranged in multiples of 72 kilometers per second. Thus he concludes that these are nonvelocity red shifts, and he posits a static universe.

Summarizing his work, Tifft makes the following remarks in the *Astrophysical Journal*:

> The entire set of concepts [developed in these papers] is internally self-consistent and permits predictions which the conventional view does not even suggest. The predictions made have been verified in virtually all cases and offer alternatives to some very puzzling astrophysical problems: the mass discrepancy problem for galaxies, and stellar rotational peculiarities, to name two major ones. Although not discussed specifically in these papers, the origin and evolution of galaxies by collapse are also untenable, as are most of the cosmological concepts based on the "expanding" universe. In view of all the implications which inevitably follow from the discrete red shift hypothesis, it is not surprising that the idea has met extreme resistance. Nevertheless, a set of intimately related significant correlations involving a massive amount of data exists. Showing that the discrete red shift concept is inconsistent with the "expanding universe" or even general relativity or quantum electrodynamics will not eliminate or explain the correlations! [TF5, p. 390]

As we can see from this statement, Tifft's conclusions have not met with a favorable reception in the community of astronomers and astrophysicists. Indeed, they have been greeted largely with a barrier of icy silence. However, Halton Arp independently confirms some of Tifft's findings, and this in turn lends greater weight to Arp's own anomalous observations.

One of Arp's observations is that in groups of galaxies, one member is generally brighter and bigger. This galaxy tends to have a lower red shift than its smaller companion galaxies. Arp suggests the galaxies are all in the same region, at the same general distance from us; therefore the red shifts are not giving velocity effects and distances but indicate something else.

Let us carefully consider the reasoning that leads Arp to this conclusion. One possibility is that the large, bright galaxy is nearby and just happens to be projected against a background of galaxies that are smaller and dimmer because of distance. These galaxies would have larger red shifts as a result of the expansion of the universe.

However, Arp argues that this explanation overlooks the fact that the clusters of galaxies are well defined and that such well-defined clusters cover a small percentage of the sky. It is therefore unlikely that many such clusters should just happen to have a bright foreground galaxy projected in front of them.

As we have already pointed out, Arp believes that galaxies can be ejected from a parent galaxy. What if the relative red shifts of the smaller galaxies are due to their being ejected from the larger parent galaxy in a direction pointing away from us? The problem here is that in this case we would expect some of the smaller galaxies to be ejected in our direction. These would exhibit relative blue shifts, contrary to Arp's observations.

But here is Arp's clinching argument: Not only do these smaller galaxies have positive red shifts relative to their parent galaxies, but these red shifts are quantized, just the way Tifft indicates they should be in his studies. Arp finds peaks at 70, 140, and 210 kps; this agrees well with Tifft's findings of quantization in multiples of 72 kps. As we have seen, this means that they are nonvelocity red shifts. And the fact that the quantization is in relation to the dominant galaxy in the group indicates there is some physical association. Why would the quantization be there if the association is simply coincidental? The fact that it is there indicates that the association is real.

So here we have an example in which we see dim galaxies with high red shifts close to bright galaxies with lesser red shifts, although standard cosmological theory says they should be vastly further away. This raises questions not only about the interpretation of red shifts, but also about the whole procedure of calculating distance according to brightness.

One of Arp's peaks for red-shift differences among groups of galaxies is in the range of 138–144 kilometers per second. This extremely narrow range is highly significant. If these groups are involved in orbital motion, we would expect to find them at different points in their orbits—some galaxies should be coming toward us, while others should be moving away from us. Thus we would expect a much greater spread in velocities than the 6 kps range found in this peak.

As Arp puts it, "The really startling and difficult aspect of the quantization into very narrow peaks is the small latitude it allows for the true orbital or peculiar velocity" (AR2, p. 110). This suggests that the orbital velocities, if present, are very small, too small for the galaxies to be actually orbiting each other according to present physical laws and estimates for the masses of the galaxies. Tifft's ideas about the need for new laws of gravity seem to be confirmed.

Geoffrey Burbidge has summed up the evidence for anomalous red shifts by saying,

> I believe that however much many astronomers wish to disregard the evidence by insisting that the statistical arguments are not very good, or by taking the approach that absence of understanding is an argument against the existence of the effect, it is there and many basic ideas have to be revised.
>
> A revolution is upon us whether or not we like it [BR2, p. 103].

What we see from this evidence is that an established model of the universe, built up from years of painstaking scientific work, can be practically demolished by closer scrutiny of the observational data on which it is based. In the end we come back to the observation made by Śukadeva Gosvāmī in the beginning of his description of the universe:

> My dear King, there is no limit to the expansion of the Supreme Personality of Godhead's material energy. This material world is a transformation of the material qualities, . . . yet no one could possibly explain it perfectly, even in a lifetime as long as that of Brahmā. No one in the material world is perfect, and an imperfect person could not describe this material universe accurately, even after continued speculation [SB 5.16.4].

Questions and Answers

Here we give brief answers to a number of common questions about Vedic cosmology. We also indicate sections in the preceding chapters where more detailed answers are given.

Q: The Vedic literature says the moon is higher than the sun. How can this be?
A: In Chapter 22 of the Fifth Canto, the heights of the planets above the earth are given, and it is stated that the moon is 100,000 *yojanas* above the rays of the sun. In this chapter, the word "above" means "above the plane of Bhū-maṇḍala." It does not refer to distance measured radially from the surface of the earth globe. In Section 4.b we show that if the plane of Bhū-maṇḍala corresponds to the plane of the ecliptic, then it indeed makes sense to say that the moon is higher than the sun relative to Bhū-maṇḍala. This does not mean that the moon is farther from the earth globe than the sun.

For example, if point A is in a plane, B is 1,000 miles above the plane, and C is 2,000 miles above the plane, we cannot necessarily conclude that C is further from A than B is.

Q: In SB 8.10.38p, Śrīla Prabhupāda says, "The sun is supposed to be 93,000,000 miles above the surface of the earth, and from the *Śrīmad-Bhāgavatam* we understand that the moon is 1,600,000 miles above the sun. Therefore the distance between the earth and the moon would be about 95,000,000 miles." Doesn't this plainly say that the moon is farther from the earth than the sun?
A: In the purport to Text 9 at the end of Chapter 23 of the Fifth Canto Śrīla Prabhupāda says, "The distance from the sun to the earth is 100,000 *yojanas*." At 8 miles per *yojana*, this comes to 800,000 miles. We suggest that when Śrīla Prabhupāda cites the modern Western earth-sun distance of 93,000,000 miles, he is simply making the point that if you put together the *Bhāgavatam* and modern astronomy you get a contradictory picture. His conclusion is that one should simply accept the Vedic version, and he was not interested in personally delving into astronomical arguments in detail.

Q: What is your justification for going into these arguments in detail?

A: Śrīla Prabhupāda ordered some of his disciples to do this for the sake of preaching. In a letter to Svarūpa Dāmodara dāsa dated April 27, 1976, Śrīla Prabhupāda said, "Now our Ph.D.'s must collaborate and study the 5th Canto to make a model for building the Vedic Planetarium. . . . So now all you Ph.D.'s must carefully study the details of the 5th Canto and make a working model of the universe. If we can explain the passing seasons, eclipses, phases of the moon, passing of day and night, etc., then it will be very powerful propaganda." In this regard, he specifically mentioned Svarūpa Dāmodara dāsa, Sadāpūta dāsa, and Mādhava dāsa in a letter to Dr. Wolf-Rottkay dated October 14, 1976.

Q: If the distance from the earth to the sun is 800,000 miles, how can this be reconciled with modern astronomy?

A: This distance is relative to the plane of Bhū-maṇḍala. The distance from the center of Jambūdvīpa to the orbit of the sun around Mānasottara Mountain is 15,750,000 *yojanas* according to the dimensions given in the Fifth Canto. This distance lies *in* the plane of Bhū-maṇḍala and comes to 126,000,000 miles at 8 miles per *yojana* and 78,750,000 miles at 5 miles per *yojana*. Since values for the *yojana* ranging from 5 to 8 miles have been used in India, this distance is compatible with the modern earth-sun distance of 93,000,000 miles.

Q: Using radar and lasers, scientists have recently obtained very accurate estimates of the earth-moon distance. This distance is about 238,000 miles. How do you reconcile this with Vedic calculations?

A: According to the *Sūrya-siddhānta*, the distance from the earth globe to the moon is about 258,000 miles (see Section 1.e). This is in reasonable agreement with the modern value.

Q: If the moon is 258,000 miles from the earth globe, then how can it be 100,000 *yojanas* above the sun? This seems hard to understand, even if the latter distance is relative to the plane of Bhū-maṇḍala.

A: This question is answered in detail in Section 4.b, and the reader should specifically study Tables 8 and 9 in that section. Briefly, we propose the following: The heights of the planets from Bhū-maṇḍala correspond to the maximum heights of the planets from the plane of the ecliptic in the visible solar system. This correspondence is approximate because the Fifth Canto gives the viewpoint of the demigods, whereas in modern astronomy and the *jyotiṣa śāstra* the viewpoint is that of ordinary humans.

In summary, we propose that the Fifth Canto description of the universe is broadly compatible with what we see. The differences are due to the difference in viewpoint between human beings and demigods. Thus, from the higher-dimensional perspective of a demigod, Bhū-maṇḍala should be directly visible, and the relative positions of Bhū-maṇḍala, the sun, and the moon should appear as described in the Fifth Canto.

Q: How are we to make sense of the enormous mountains described in the Fifth Canto? Some of them, including the Himalayas, are said to be 80,000 miles high.
A: One might well doubt that even a scientifically uneducated person in ancient India would have thought that the Himalaya Mountains of our ordinary experience are 80,000 miles high. After all, such persons traditionally made pilgrimages to Badarikāśrama on foot. We suggest that the cosmic mountains of the Fifth Canto are higher-dimensional; they are real, but to see them it is necessary to develop the sensory powers of the demigods and great yogis. This is the traditional understanding, although words such as "higher-dimensional" are not used, and descriptions are made in a matter-of-fact way from the viewpoint of demigods and other great personalities (such as the Pāṇḍavas).

Śrīla Prabhupāda has said that modern scientists are "hardly conversant with the planet on which we are now living" (SB 5.20.37p). If our ordinary three-dimensional continuum is the total reality, then this statement would seem to be wrong. In Section 3.b.4, however, we give Vedic evidence showing that this three-dimensional world links up with higher-dimensional realms.

Q: If the Garbhodaka Ocean fills half the universe, where is it, and why don't we see it?
A: The Garbhodaka Ocean is beneath Bhū-maṇḍala. Thus its location corresponds to the region of the celestial sphere south of the great circle marked by Bhū-maṇḍala. We have argued that this should be either the southern celestial hemisphere or the region to the south of the ecliptic (see Section 3.d). The Garbhodaka Ocean is also higher-dimensional.

Q: Isn't it true that there are fewer stars in the southern celestial hemisphere than in the northern celestial hemisphere? Isn't this because we are looking down on Bhū-maṇḍala from the earth?
A: A study of standard star charts shows that the number of stars visible in the southern celestial hemisphere is essentially the same as the number visible in the northern celestial hemisphere. (See Figures 11 and 12.)

Q: What was Śrīla Prabhupāda's position on the moon flight? There seems to be some ambiguity in his statements about this topic.
A: Śrīla Prabhupāda offered a number of tentative explanations as to what might have actually transpired on the moon flight, but his main point was that the astronauts could not have visited Candraloka, since they did not reach the civilization of the demigods that exists there. To put the matter in another way, if the moon is really nothing more than a lifeless desert, as scientists maintain, then the Vedic literatures describing Candraloka must be wrong. This topic is discussed in Section 6.c.1.

Q: What about the argument that the moon flights were faked by the U.S. government? A case for this is made in the book, *We Never Went to the Moon*, by Bill Kaysing.

A: Although this book makes some interesting points, its arguments are basically speculative and circumstantial. One of Kaysing's main arguments is that Thomas Baron, a North American Aviation employee who wrote a report critical of the Apollo program, was murdered by government agents. Kaysing maintains that this was done as part of a government cover-up of the moon hoax. Unfortunately, if this is true, then it would be very dangerous to possess solid evidence proving such a cover-up. Another point made by Kaysing is that according to official reports, six Apollo flights to the moon were nearly flawless in execution. In contrast, the history of space flight before and after the Apollo program is filled with stories of failures and mechanical breakdowns. Kaysing argues that this is statistically unlikely, and cites this as evidence that the Apollo flights were faked. This argument is interesting, but certainly not conclusive.

Q: What is the justification for bringing in works of Indian mathematical astronomy, such as the *Sūrya-siddhānta* and *Siddhānta-śiromaṇi*?

A: Śrīla Prabhupāda follows Śrīla Bhaktisiddhānta Sarasvatī by citing these works, which are called *jyotiṣa śāstra*. He does so because Śrīla Bhaktisiddhānta Sarasvatī cited these works in his writings. Śrīla Bhaktisiddhānta gives direct quotations and says nothing indicating that the works are wrong in any way. Also, the *jyotiṣa śāstras* are cited by other Vaiṣṇava commentators on the *Bhāgavatam*. See Chapter 1.

Q: Śrīla Prabhupāda refers to the earth as a globe, and Śrīla Bhaktisiddhānta Sarasvatī Ṭhākura made references to *Sūrya-siddhānta* and other *jyotiṣa śāstras* that describe the earth as a globe. But wasn't this an innovation introduced by Śrīla Bhaktisiddhānta in response to modern astronomy?

A: This is not an innovation introduced by Śrīla Bhaktisiddhānta. Earlier commentators on the *Śrīmad-Bhāgavatam* make reference to the *jyotiṣa śāstras*, including the *Sūrya-siddhānta*. One example is Vaṁśīdhara, who was living in A.D. 1642, before the time that Western science made a large impact on India (see Appendix 1).

Q: Does this mean that we have to accept the *jyotiṣa śāstras* as absolute truth on the level of the *Śrīmad-Bhāgavatam*?

A: No. The *Śrīmad-Bhāgavatam* is the spotless *Purāṇa*, containing pure knowledge of the Supreme Personality of Godhead. The *jyotiṣa śāstras* are handbooks for the execution of astronomical calculations. The *Bhāgavatam* presents the world from a transcendental perspective, or at least gives the perspective of great personalities involved in Kṛṣṇa's pastimes. The *jyotiṣa śāstras* deal with the motions of planets as seen by ordinary human beings. However, the *jyotiṣa śāstras* do form a valid part of Vedic tradition, and their calculations are mentioned by Śrīla Prabhupāda in various places.

Q: Scholars say the calculations given in the *jyotiṣa śāstras* were borrowed from the Greeks in the early centuries of the Christian era. How do we deal with this?

A: Western scholars maintain that all the Vedic literature is relatively recent. However, their methods are speculative, and they are not free of ethnic and religious bias. In Appendix 2, we show the baseless nature of some of their arguments.

Q: Some have said that the description of the universe in the Fifth Canto is allegorical and that *Bhāgavatam* commentators have declared this. For example, Bhaktivinoda Ṭhākura has said that the descriptions of hell are allegorical. Why don't you just accept the Fifth Canto as an allegory and leave it at that?

A: It would indeed make things easier if we could simply accept the description of the universe in the Fifth Canto as an allegory. But in good conscience we cannot do so. Let us carefully consider the reasons for this.

First of all, consider the statements of Bhaktivinoda Ṭhākura about descriptions of hell in the *Bhāgavatam*. In *The Bhāgavata* he writes, "In some of the chapters we meet with descriptions of these hells and heavens, and accounts of curious tales, but we have been warned somewhere in the book not to accept them as real facts, but as inventions to overawe the wicked and improve the simple and ignorant. The *Bhāgavata* certainly tells us of a state of reward and punishment in the future according to deeds in our present situation. All poetic inventions besides this spiritual fact have been described as statements borrowed from other works."

According to this passage, not only the hells but also the material heavens are dismissed as poetic inventions. But if the heavens are inventions, what can one say about their inhabitants, such as Indra? If Indra is also imaginary, then how are we to understand the story of the lifting of Govardhana Hill? This must also be imaginary, and we are led to an allegorical interpretation of Kṛṣṇa's pastimes.

In *The Bhāgavata* Bhaktivinoda Ṭhākura is indeed introducing the *Bhāgavatam* in this way. We would suggest that he is doing this in accordance with time and circumstances. He describes his readers in the following words: "When we were in college, reading the philosophical works of the West, . . . we had a real hatred towards the *Bhāgavata*. That great work looked like a repository of wicked and stupid ideas scarcely adapted to the nineteenth century, and we hated to hear any arguments in its favor." In order to sidestep the strong prejudices of readers trained by the British in Western thinking, Bhaktivinoda Ṭhākura is presenting the *Bhāgavatam* as allegorical, but we would suggest that this is not his final conclusion.

Śrīla Prabhupāda has explained that the Vedic literatures should be understood in terms of *mukhya-vṛtti*, or direct meaning, rather than *gauṇa-vṛtti*, or indirect meaning. He has also said, "Sometimes, however, as a matter of necessity, Vedic literature is described in terms of the *lakṣaṇā-vṛtti* or *gauṇa-vṛtti*, but one should not accept such explanations as permanent truths" (CC AL 7.110p).

Bhaktivinoda Ṭhākura was reviving Vaiṣṇavism at a time when it had almost completely disappeared because of internal deviations and Western propaganda, and he may have concluded that an allegorical presentation was necessary under those circumstances.

In establishing the foundations of Vaiṣṇavism in the West, Śrīla Prabhupāda stressed the importance of the direct interpretation of *śāstra*. He writes, "Considering the different situation of different planets and also time and circumstances, there is nothing wonderful in the stories of the *Purāṇas*, nor are they imaginary. . . . We should not, therefore, reject the stories and histories of the *Purāṇas* as imaginary. The great *ṛṣis* like Vyāsa had no business putting some imaginary stories in their literatures" (SB 1.3.41p).

But could the description of the universe in the Fifth Canto be an allegory like the story of King Purañjana? Śrīla Prabhupāda makes many statements indicating that this not so. For example, he says that "we can understand that the sky and its various planets were studied long, long before *Śrīmad-Bhāgavatam* was compiled. . . . The location of the various planetary systems was not unknown to the sages who flourished in the Vedic age" (SB 5.16.1p). He also says, "The measurements given herein, such as 10,000 *yojanas* or 100,000 *yojanas*, should be considered correct because they have been given by Śukadeva Gosvāmī" (SB 5.16.10p).

In this book we have therefore tried to show that the Fifth Canto is giving a reasonable picture of the universe consistent with (1) transcendental Vedic philosophy, (2) the tradition of Vedic mathematical astronomy, and (3) our imperfect sense data.

Q: To my knowledge, Śrīla Prabhupāda never hinted at explanations of other dimensions; he always seemed to emphasize accepting it as it is written. If these ideas are right, why didn't Śrīla Prabhupāda save us a lot of trouble by bringing them out years ago?

A: The Vedic literature does not explicitly refer to the concept of higher-dimensional space, as far as I am aware. This idea is borrowed from modern mathematics. However, the Vedic literature does refer *implicitly* to higher-dimensional space, and therefore it is justifiable to use this idea to clarify the Vedic description of the universe.

For example, in the description of Lord Brahmā's visit to Kṛṣṇa in Dvārakā, it is stated that millions of Brahmās from other universes came to visit Kṛṣṇa. However, each Brahmā remained within his own jurisdiction, and apart from our Brahmā, each thought he was alone with Kṛṣṇa. Thus Kṛṣṇa was in many universes at once, and our Brahmā could also simultaneously see different Brahmās visiting Kṛṣṇa in all of these universes. This is impossible in three dimensions; it illustrates the implicit higher-dimensional nature of the Vedic conception of space (see Chapter 2).

Q: If we could visit the moon, would the inhabitants be visible to us or invisible?
A: Śrīla Prabhupāda has said "almost invisible" (see Section 6.c.1).
Q: Couldn't it be that denizens of higher planets are invisible to us simply because they have subtle bodies? Why bring in the idea of higher-dimensional worlds?
A: The clothes, food, dwellings, airplanes, and other paraphernalia of the demigods must be just as invisible to us as the demigods themselves. (Imagine what it would be like to see a suit of clothes being worn by an invisible demigod!) In other words, the demigods live in a complete world that is invisible to us but perfectly visible to them. They can travel to our world since they are endowed with suitable mystic powers, and advanced yogis can travel to their world. However, humans with ordinary senses cannot perceive the demigods or their gardens and cities. This sums up what we *mean* by a higher-dimensional world.

If we use the word "subtle," we should realize that we are speaking of a complete subtle world that looks perfectly substantial to the persons living in it, just as our world looks substantial to us. The worlds of the demigods should be contrasted with the situation of a ghost, who is stranded in our own continuum in a subtle form, but is unable to enjoy it.
Q: These higher-dimensional worlds may be normally inaccesible to us, but if they are actually real, shouldn't there be some empirical evidence of them? Do we just have to accept this whole incredible story on blind faith?
A: There is abundant empirical evidence of higher-dimensional worlds, and such evidence has been well known in practically all human cultures since time immemorial. Our modern scientific culture is an exception in this regard.

In Chapter 5 we briefly discuss some empirical evidence taken from non-Vedic sources.
Q: But isn't this empirical evidence imperfect?
A: Empirical evidence is always imperfect. One may accept the version of *śāstra* according to the descending process, or one can turn to the empirical process with all its imperfections. Of course, Śrīla Prabhupāda advocated the descending process.
Q: There are places in the *Śrīmad-Bhāgavatam* where it is said that the coverings of the universe begin with water. Since this is clear water, and the farther coverings are transparent, it should be possible for us to see the suns of other universes. Couldn't these be the stars we see in the sky at night?
A: In SB 5.21.11p, Śrīla Prabhupāda says, "The Western theory that all luminaries in the sky are different suns is not confirmed in the Vedic literature. Nor can we assume that these luminaries are the suns of other universes, for each universe is covered by various layers of material elements, and therefore although the universes are clustered together, we cannot see from one universe

to another. In other words, whatever we see is within this one universe."

In Section 6.d it is shown that the coverings of the universe are listed four times in the *Bhāgavatam* as beginning with earth. We suggest that when Śrīla Prabhupāda mentions water or fire first, he is giving a partial list of the coverings.

Q: In SB 5.16.5, Jambūdvīpa is described as having a length and breadth of one million *yojanas*, yet in SB 5.16.7, it is described as having a width that is the same as Sumeru's height, namely 100,000 *yojanas*. This seems contradictory. In SB 5.16.7, Sumeru's width is stated to be 32,000 *yojanas* at its summit, and in SB 5.16.28, the township of Brahmā has sides that extend for ten million *yojanas*. Does Brahmapurī hang way out over the edge of Sumeru?

A: The correct diameter of Jambūdvīpa is 100,000 *yojanas*, since this figure agrees with all the other dimensions mentioned in the Fifth Canto. Likewise, the width of Sumeru at its summit is 32,000 *yojanas*. We do not know the explanation for the other figures.

Q: What can be said in general about such apparent contradictions in the *Bhāgavatam*? Does it mean that we should not have faith in it as a source of absolute truth?

A: Certainly it would not be justifiable to draw such conclusions from minor discrepancies. In many cases the discrepancy may have an explanation that we cannot guess because we have too little information. For example, in the Third Canto, two boar incarnations of Lord Viṣṇu are mentioned. In certain verses there appears to be some ambiguity in the description of these incarnations, and Śrīla Prabhupāda cites Śrīla Viśvanātha Cakravartī as saying that "the sage Maitreya amalgamated both the boar incarnations in different devastations and summarized them in his description to Vidura" (SB 3.13.31p). Without this information from Śrīla Viśvanātha Cakravartī, we might find it difficult to resolve the apparent contradictions in the story of Lord Varāha.

We suggest that some of the apparent contradictions discussed in Section 3.d may have a similar explanation.

Q: SB 5.17.6 places Bhadrāśva-varṣa west of Mount Meru, and SB 5.17.7 says the same thing about Ketumāla-varṣa. How do you resolve this contradiction?

A: If we look at the Sanskrit texts of these verses, we find that Bhadrāśva and Ketumāla varṣas are on opposite sides of Mount Meru. Careful inspection of SB 5.16.10 shows that Bhadrāśva-varṣa is to the east of Mount Meru, since its boundary mountain is Mount Gandhamādana.

Q: SB 5.24.2 says that the moon is twice as big as the sun, and Rāhu is three times as big. The purport says that Rāhu is four times as big as the sun. How do you explain this?

A: This is another case of an apparent contradiction. Since we have practically no information, we cannot make a definite statement. But it is possible that the large sizes of the moon and Rāhu may have to do with the higher-dimensional aspects of these planets.

The *Sūrya-siddhānta* gives a diameter of 2,400 miles for the moon. This is close to the modern figure (see Section 1.e).

Q: I have heard that all of the planets are in the stem of the lotus from which Brahmā took birth. How can that be?

A: This is stated in SB 1.3.2p. Since the planetary systems are distributed throughout the universal globe, it must be that the stem encompasses everything within this globe. We should note that the standard pictures we see of Brahmā sitting on the lotus flower are three-dimensional representations of a scene that cannot be seen using our ordinary senses. Although the pictures show the lotus stem emerging from the navel of Garbhodakaśāyī Viṣṇu, Brahmā himself was unable to locate the origin of the stem. Thus, part of the scene was beyond the senses of Brahmā, and so it is certainly beyond the reach of our senses. We also note that the planetary systems were created by Brahmā from the lotus (SB 3.10.7–8). This suggests that these systems were produced by transforming the substance of the lotus.

Q: The *Bhāgavatam* says that Rāhu causes the eclipses of the sun and moon. How can this be reconciled with modern science?

A: The *jyotiṣa śāstras*, such as *Sūrya-siddhānta*, give the same explanation of solar and lunar eclipses as modern science. These *śāstras* also describe the orbit of Rāhu (and Ketu) and point out that eclipses occur only when one of these two planets is aligned with either the sun and the moon or the earth's shadow and the moon (see Section 4.e). Some will maintain that this account was devised centuries ago to reconcile Vedic *śāstras* with Greek astronomy. But this is sheer speculation.

Q: What can be said about the precession of the equinoxes and the consequent displacement of the polestar?

A: The phenomenon of precession is described in the *jyotiṣa śāstras*. We discuss this topic in Section 4.f.

Q: The *Bhāgavatam* says that the diameter of the universe is 4 billion miles. This is much too small to accommodate even the solar system, what to speak of the stars and galaxies. How can the *Bhāgavatam* be correct?

A: Śrīla Prabhupāda, citing Śrīla Bhaktisiddhānta Sarasvatī, also gives a figure of 18,712,069,200,000,000 *yojanas* for the circumference of the universe (or half the circumference) (CC ML 21.84p). He also says that "scientists calculate that if one could travel at the speed of light, it would take forty thousand years to reach the highest planet of this material world" (SB 3.15.26p).

We suggest that cosmic distances may appear different to observers endowed with different levels of consciousness. We also suggest that the laws governing distance and time may not be the same in outer regions of the universe as they are here on the earth (see Sections 1.f and 4.c).

Q: Scientists in the twentieth century have amassed a huge amount of information about distant stars and galaxies. How can you lightly suggest that it may be seriously wrong?

A: In Chapter 7 we discuss some of the latest findings of modern cosmology. There is abundant evidence in standard scientific journals to show that modern cosmological theories have serious defects.

Q: The scientists say that spectroscopic studies show that the stars are incandescent bodies and not mere reflectors of light. They also say that the stars are typically as powerful or more powerful than the sun, and they have worked out in detail the thermonuclear reactions that sustain stellar radiation. How can this be reconciled with the Vedic version?

A: This is discussed in Section 6.e. Briefly, we suggest that stars may well give off their own light. However, the Vedic literature indicates that they cannot be independent suns. The highly detailed scientific theories about stars may well be wrong in many respects. After all, these theories are based entirely on the interpretation of starlight. Their underlying logic is: This model seems to fit the data, and therefore it should be accepted and taught to students. Chapter 7 shows some of the pitfalls of this approach.

APPENDIX 1

VAṀŚĪDHARA ON BHŪ-MAṆḌALA AND THE EARTH GLOBE

In this appendix we will discuss a commentary on verse 5.20.38 of the Fifth Canto of *Śrīmad-Bhāgavatam*, written in the 17th century by Vaṁśīdhara. It is included in his *Bhāgavatam* commentary, entitled *Bhavārtha-dīpikā-prakāśa*, which appears in the compilation of eleven *Bhāgavatam* commentaries used by Śrīla Prabhupāda when writing his purports. (*Bhavārtha-dīpikā* is the title of Śrīdhara Svāmī's commentary.) This commentary explicitly discusses the relationship between Bhū-maṇḍala, as described in the Fifth Canto, and the small earth globe of our experience.

We will summarize his commentary here, since it sheds some light on how Vedic cosmology and astronomy were regarded by Vaiṣṇavas in India before the widespread introduction of modern Western ideas. It shows that the cosmology of the Fifth Canto was controversial during the period of the 1600s, when Vaṁśīdhara was active. It also shows that the astronomical literature known as *jyotiṣa śāstra* was accepted as valid by Vaiṣṇavas, and it discusses the apparent contradiction that exists between the cosmology of the Fifth Canto and this system of astronomy.

Vaṁśīdhara tries to resolve this contradiction, and we should state clearly here that we do not think that his analysis is entirely correct. However, our own understanding is certainly far from perfect. At the present time in history, when much ancient Vedic knowledge has been lost, it is difficult to reconstruct many important aspects of ancient astronomical science. Thus it is best for us to carefully consider the information that is available to us and see what insights may gradually emerge.

We begin by quoting Śrīla Prabhupāda's translation of SB 5.20.38:

> Learned scholars who are free from mistakes, illusions, and propensities to cheat have thus described the planetary systems and their particular symptoms, measurements, and locations. With great deliberation, they have established the truth that the distance between Sumeru and the mountain known as Lokāloka is one fourth of the diameter of the universe—or, in other words, 125,000,000 *yojanas* [1 billion miles].

The first section of Vaṁśīdhara's commentary on this verse is also stated in the commentary of Viśvanātha Cakravartī Ṭhākura, and Śrīla Prabhupāda reproduces it in Sanskrit in his purport. This section begins by pointing out that the word *bhū-golasya* in the verse means "of the egglike sphere connected with the earth." This egglike sphere is the inner shell of the universe and has a diameter of 500 million *yojanas*. According to the Fifth Canto, the earth has the same diameter, and thus it should touch the universal shell on all sides. However, Vaṁśīdhara and Viśvanātha Cakravartī Ṭhākura point out that the earth actually should be assigned a diameter of 496,600,000 *yojanas*. This figure is twice the sum of the following distances in lakhs (or 100,000s) of *yojanas*: (1) 157.5 from Mount Meru to Mānasottara Mountain, (2) 96 from there to the outer shore of the clear-water ocean, (3) 157.5 for the width of the inhabited land, (4) 822 [or 4 × 157.5 + 2 × 96] for the width of the golden land bounded by Lokāloka Mountain, and (5) 1,250 for Aloka-varṣa, which lies beyond Lokāloka Mountain.

As a result of this revised value for the diameter of the earth, there is a gap of 17 lakhs of *yojanas* between the earth and the universal shell on all sides. The commentators point out that this gap makes it possible for the earth to move within the universal shell. This makes it meaningful for the earth to be supported by Ananta Śeṣa, and it also allows the earth to be immersed in the Garbhodaka Ocean in the Cākṣuṣa *manvantara* and be lifted by Lord Varāha. We can understand from this that the earth lifted by Lord Varāha is the complete Bhūmaṇḍala of approximately 500 million *yojanas* in diameter and not the small earth globe of our experience (see also Chapter 3.c).

In the next section of his commentary, Vaṁśīdhara confronts the apparent conflict between the size of the earth given in the *Bhāgavatam* and the size given in the *jyotiṣa śāstra*. From *jyotiṣa śāstra*, he cites a value of 4,967 *yojanas* for the circumference of the earth globe. In fact, this figure is given in verse 3.52 of the *Siddhānta-śiromaṇi* of Bhāskarācārya, along with a value of 1,581 1⁄24 *yojanas* for the earth's diameter (SSB1, p. 122). As we have pointed out in Chapter 1, this agrees closely with our present figure for the circumference of the earth, using 5 miles per *yojana*. Vaṁśīdhara does not indicate that he thinks this figure is wrong. Rather, he accepts it without question and suggests various ways of reconciling it with Purāṇic cosmology. These are as follows:

(1) Śrī Nīlakaṇṭha, in his commentary on the *Bhīṣma-parva* of the *Mahābhārata*, gives a description of Jambūparvan as a square with its diagonals oriented north-south and east-west. This square is also described as a lotus with a perimeter of 18,600 *yojanas* and an inner diameter of 3,300 *yojanas*. Nīlakaṇṭha argues that since one side of such a square is 4,650 *yojanas* in length, the size of Jambūparvan agrees in a crude "order of magnitude" fashion with the size of the earth globe given in the *jyotiṣa śāstra*. In addition, since Bhāratavarṣa corresponds to the southern part of the Jambūparvan square and is bounded by the Himalaya Mountains on the north, it follows that Bhārata-varṣa must be triangular. This agrees with ordinary experience, whereas the idea that Bhārata-varṣa is bow-shaped does not. (The idea that Bhārata-varṣa is bow-shaped fol-

lows from the Purāṇic description of Bhārata-varṣa as the southern part of the disc of Jambūdvīpa.)

Here, a rough agreement is achieved between Jambūparvan and the earth globe of the *jyotiṣa śāstra*, but at the same time a contradiction is introduced between this account of Jambūparvan and the Jambūdvīpa of the *Śrīmad-Bhāgavatam*. In Chapter 3 we argued that Jambūdvīpa is inherently higher-dimensional and that it can be seen in different ways, depending on one's level of consciousness.

(2) Rather than introduce this idea, Vaṁśīdhara suggests that the apparent contradiction posed by the 500-million-*yojana* earth diameter in the *Purāṇas* can be resolved by resorting to the principle of inexplicability (*anirvacanīya-vāda*). According to this principle, "One should not try to establish by logic or argument those things that are beyond imagination."

(3) Then he suggests that whatever measure is mentioned in the *Purāṇas*, one should take 1⁄20 of that, and thus in place of 500 million one should accept 25 million *yojanas* as the measure of the earth. In place of one lakh, one should accept 5,000 *yojanas* as the diameter of Jambūdvīpa, and in place of 9,000, one should accept 450 *yojanas* as the measure of Bhārata-varṣa. We note that 9,000 *yojanas* is the width of Bhārata-varṣa from north to south, according to the *Bhāgavatam*. (This distance is roughly 1,600 miles on a modern map.)

(4) To justify these reductions in scale, Vaṁśīdhara observes that the *yojana* is defined on the basis of the human body. Thus a *yojana* is 32,000 *hastas* (or cubits), and a *hasta* is 24 finger-widths. Also, a *hasta* can be defined as one fifth the height of a man standing with his arms stretched up. As the bodies of infants, children, and adults vary greatly in size, so the *yojana* also varies, and in this way one can explain differences between various estimates of distance.

After offering these arguments, Vaṁśīdhara sums up his position in the following words:

> Not indeed has the *jyotiṣa śāstra*, or science of luminaries, started contravening the Purāṇic statement that "Vyāsa is Nārāyaṇa himself." Nor could Vyāsa also have proceeded in contravention of the science of luminaries, which is the very eye of *Veda*, as expressed in the statement "Astronomy is declared to be the eye [of the *Veda*]." Therefore, at different places, statements as to *yojanas* may be inferred to mean these various measures of finger, hand, etc. Moreover, it appears that Vyāsadeva himself speaks contrary to astronomy, as it were, in order to curb the tendency on the part of *asuras* toward the study of the *śāstras*. But truly speaking, he is not doing so. Otherwise, it may be contemplated by the well-intentioned that there would be darkness [i.e., ignorance] as to *Veda* on the part of Vyāsa. This is the proper understanding.

The statement that astronomy is the eye of the *Veda* may refer to verse 1.4 of the *Nārada-saṁhitā*: "The excellent science of astronomy comprising *siddhānta*, *saṁhitā*, and *horā* as its three branches is the clear eye of the *Vedas*" (BJS, p. xxvi). "*Siddhānta*," of course, refers to works such as the *Sūrya-siddhānta*.

Vaṁśīdhara is not satisfied with the explanations that he has given thus far. He goes on to give a sharper formulation of the basic problem:

> Well, then how can one explain the contradiction between the *Bhāgavata* and the *jyotiṣa śāstra*, or astronomical science? In the *Bhāgavata*, Jambudvīpa is said to measure 100,000 *yojanas*, whereas the astronomical science states the entire earth to be measuring only 5,000 *yojanas*. The solution is given in the *Goladarśa*. According to that text, some brief explanation is given below:
>
> The earth has two forms. One is the particular [*viśeṣa*] form of big measure, and the other is the smaller, spherical form given in the *jyotiṣa śāstra*. In this regard Parīkṣit asked Śrī Suka, the great *yogin*, and he replied [in SB 5.16.4]: "We shall explain the particular description of *bhūgola* by name, form, measure, and characteristics." In the *jyotiṣa śāstra* the word *bhūgola* refers to the earth as an egg of clay, and the word *viśeṣa*, or "particular," refers to the round golden egg described in the *Purāṇas*.

Here a clear distinction is made between the earth of our experience, which is described in the *jyotiṣa śāstra* as having a diameter of 5,000 *yojanas*, and another earth "of big measure," described by Śukadeva Gosvāmī. The figure of 5,000 *yojanas* is a simple approximation of the 4,967-*yojana* diameter of the earth given in the *Siddhānta-śiromaṇi*. The earth of big measure can be thought of as either the spherical inner shell of the universe or the disc-shaped Bhū-maṇḍala. Both are made of the earth element, and both have a diameter of about 500 million *yojanas*. However, since Bhū-maṇḍala contains the seven *dvīpas* and oceans, it is clear that the big earth really should correspond to Bhū-maṇḍala.

Vaṁśīdhara then cites a number of verses from the *Bhāgavatam* to illustrate his point concerning the existence of two earths:

> It is stated in the Second Canto, the *yoginīryāna* [SB 2.2.28], "then reaching the particular, or *viśeṣa*, one becomes fearless." It is also said in the Fifth Canto [5.20.35], "There is another land, made of gold, with a mirrorlike surface," etc. Also, in the Third Canto [3.26.52] it is stated, "This universal egg is called particular or manifest [*viśeṣa*], with tenfold increasing coverings." In the Fifth Canto [5.25.2]: "This great universe [*kṣiti-maṇḍalam*, or earth-*maṇḍala*], situated on one of Lord Anantadeva's thousands of hoods, appears just like a white mustard seed." Thus, by the illustration of a mustard seed it is known to be spherical. It is also said in the *Kardama-vihāra* [3.23.43], "After showing his wife the globe [*golam*] of the universe [*bhuvaḥ*] and its different arrangements, full of many wonders, the great *yogī* Kardama Muni returned to his own hermitage." In the Tenth Canto [10.8.37] it is stated, "She [Yaśodā] saw within His mouth all moving and nonmoving entities, outer space, and all directions, along with mountains,

> islands, the surface of the earth [*bhū-golam*], the blowing wind, fire, the moon, and the stars." By such proofs one should accept that there are two earths.

Some of these verses illustrate the meaning of the term *viśeṣa* used by Śukadeva Gosvāmī to describe the universe. Śrīla Prabhupāda clarifies the meaning of *viśeṣa* by translating it in SB 3.26.52 as "the manifestation of material energy." The reference to *viśeṣa* in SB 2.2.28 stresses the subtle aspects of this manifested energy, since it refers to the attainment of a subtle form by a yogi who has reached Satyaloka.

Verses 5.25.2 and 3.23.43 refer to the globe of the universe, and 10.8.37 refers to mother Yaśodā's seeing the earth globe within Kṛṣṇa's mouth. Since Vamśidhara takes these texts to refer to the two earths, he is clearly thinking of the earth as having a spherical form. He does not seem to make a clear distinction between the disc-shaped Bhū-maṇḍala and the shell of the universe.

The reference to verse 5.20.35 introduces the golden land that lies within the ring of Lokāloka Mountain in Bhū-manḍala. This golden land is said to reflect light like the surface of a mirror, so that any object that falls on it cannot be seen. This turns out to be the key to Vaṁśīdhara's solution of the dilemma of the two earths. He continues,

> Well, this earth is the smaller one, so where is the other, bigger one? The answer is: The bigger one is indeed a form of reflection [*pratibimba-rūpa*] up above the orbit of the asterisms. Its measure, according to SB 5.21.19, is "the 95,100,000-yojana circumference of the sun's orbit around Bhū." Thus, roughly speaking, it comes out to be the upper portion of the orbit of the asterisms.

Here Vaṁśīdhara proposes that the small earth is the one we live on, whereas the big earth of the *Purāṇas* is a form of reflection, or *pratibimba-rūpa*. The term *bimba* means "a mirror," and *pratibimba* means "a reflection." *Bimba* can also indicate the disc of the sun or moon, and *pratibimba* can thus indicate the sun or moon reflected from water. On a more abstract level, *bimba* means "an original object," and *pratibimba* means "a counterfeit" or "an object with which the original is compared." The term "asterism" (or *nakṣatra*) means "star constellation," and the orbit of asterisms is the orbit followed by the stars as they circle the earth.

Vaṁśīdhara then explains the idea of the big earth as a reflection:

> Then how does the reflection appear, and how does it have this form? It is like this: On all sides of the earth of 5,000 *yojanas*' circumference, separated at a distance of one *yojana*, there is the fire sphere [*anala-golaḥ*]. Thereby, above the orbit of asterisms that seems small in the distance, the golden land of pure form creates a screen of light. Therein, on all sides, is the great reflection.

> From a distant place a big thing looks small, and from another place a small thing looks big. Likewise, external objects of the universe such as the moon, look small, whereas the earth-globe, which is close by, looks big.

The Sanskrit in this passage is difficult to translate. However, the general sense seems to be that the big earth is a reflection from the golden land mentioned in SB 5.20.35. There, the golden land is described as being like a mirror, and one can imagine the earth being reflected from a vast, spherical mirror centered on the earth and situated beyond the orbit of the stars. One should note, however, that the Bhū-maṇḍala described in the Fifth Canto is inhabited, and it is therefore hard to see how it can be interpreted as a reflection.

Vaṁśīdhara then discusses the fire-sphere, and also introduces a water-sphere. We have not seen any reference to these structures in the *Bhāgavatam* or in the available *jyotiṣa śāstras*. However, Vaṁśīdhara gives a reference to the water-sphere from the *Puliṣa-siddhānta* (a work that unfortunately seems to be lost):

> What is the evidence for the existence of the fire sphere, or *anala-gola*? The evidence is that from the surface of the earth up to the limit of the orbits of the planets, there are eight divisions of winds, beginning with *āvaha*, and at the conjunction of the two [the earth and the winds], there is the water-sphere, or *jala-golaḥ*. It is mentioned in the *Puliṣa-siddhānta* that "the grasslike water-sphere is at the conjunction of the earth-air [*bhu-vāta*] and the *udvaha* wind, and by it the rays of the sun and other luminaries are seen to be separated and joined together.
>
> The *Sūrya-siddhānta* [12.46] says, "Owing to closeness, the sun's rays are vehement in summer [in the *devas*' regions]." Here "closeness" and "farness" could not exist without the reflection of the fire-sphere.

The water-sphere and the fire-sphere seem to serve as specific mechanisms for reflecting and refracting light. According to the *Siddhānta-śiromaṇi* (SSB1, p. 127), the following seven winds are listed: *āvaha* (or atmosphere), *pravaha*, *udvaha*, *samvaha*, *suvaha*, *parivaha*, and *parāvaha*. The atmosphere is 12 *yojanas* thick, and the *pravaha* wind envelops the fixed stars and planets, sweeping them westward at a uniform rate. This indicates that the water-sphere must be above the stars and planets, since it is connected with the *udvaha* wind.

Vaṁśīdhara then argues that Śukadeva Gosvāmī followed Purāṇic tradition by describing the big earth and giving only brief hints of the small earth of 5,000 *yojanas*:

> In accordance with the *Purāṇas*, Śukadeva Gosvāmī has spoken of the big measure of the earth, and only suggested the small measure; thus is the contradiction avoided by some. In the same way, although there is a contradiction involving the sphere of the sky, it is removed.

As far as we can see, this seems to be a valid point. The *Bhāgavatam* generally refers to Bhū-maṇḍala when it speaks of the earth. References to Bhū-gola, or the earth-globe, generally seem to refer to the globe of the universe, and there is no specific mention of an earth-globe 5,000 *yojanas* in circumference. However, there are some references to Bhū-gola, such as SB 10.8.37, quoted above, which may refer to this earth. Also, the idea of the earth as a sphere is strongly suggested by the description in SB 5.21.8–9 of how the sun rises at a point opposite to where it sets.

However, it is hard to see why Śukadeva Gosvāmī would elaborately describe a reflection, while only indirectly hinting at the real earth. We would suggest that "the big earth" corresponds to the reality directly perceived by persons on the level of consciousness of Śukadeva Gosvāmī, while "the small earth" corresponds to the reality perceived at an ordinary level of human consciousness. The two earths are both aspects of one underlying reality, and the relation between them is higher-dimensional: it cannot be understood in terms of the bending of light in ordinary, three-dimensional space. According to this idea, both earths are reflections, in an abstract sense, of the underlying reality.

In the remainder of his commentary, Vaṁśīdhara uses the idea of reflection to interpret a number of verses in the Fifth Canto. First, he explains SB 5.21.2, where outer space, or *antarikṣa*, is compared to the empty space between two halves of a bean or a grain of wheat. In this analogy, the lower half of the bean corresponds to the hemisphere of the universe containing Bhū-maṇḍala and the Garbhodaka Ocean, and the upper half corresponds to the hemisphere containing the higher planetary systems. The space between the two halves of the bean corresponds to a thin, flat disc of space between the lower and upper hemispheres. This space, or *antarikṣa*, is bounded below by the plane of Bhū-maṇḍala and above by the parallel plane of Bhuvarloka.

After explaining this verse, Vaṁśīdhara turns to SB 5.21.3:

> SB 5.21.3 states, "In the midst of the middle region [*antarikṣa*] is the most opulent sun." This means that the water-sphere is the seeming middle of *antarikṣa*. Just as *antarikṣa* lacks a center, so also the water-sphere lacks a center (middle) due to its sphericity. Thus the sun, which goes there, is established with the form of a reflection [*pratibimba-rūpena*]. But the real sun disc [*bimba-rūpena*] is within 125,000 *yojanas* of the center of the earth.

SB 5.21.3 states that the sun is in the midst of the disc-shaped region of *antarikṣa*, between the parallel planes of Bhū-maṇḍala and Bhuvarloka (see also SB 5.20.43). However, Vaṁśīdhara argues that space can have no middle, and goes on to say that the apparent presence of the sun in mid-space is an illusion due to reflection from the water-sphere. He then states that the real sun is within 125,000 *yojanas* of the earth globe.

The Sanskrit here is terse and difficult to translate, but the import of Vaṁśīdhara's statement seems to be as follows: The earth is a small globe, and the sun orbits it at a distance of no more than 125,000 *yojanas*. A process of reflection gives the impression that it is at a much greater distance.

Śrīla Prabhupāda indicates that the height of the sun above Bhū-maṇḍala is 100,000 *yojanas* (SB 5.23.9p). However, one cannot conclude from this that the sun circles the center of the earth globe in an orbit with a radius of 100,000 *yojanas*. The reason for this is that SB 5.21.7 states that the circumference of the sun's orbit is 95,100,000 *yojanas*. Actually, Bhū-maṇḍala is a plane rather than a globe, and the *Bhāgavatam* states that the sun moves in a large orbit parallel to this plane and very close to it. In Chapter 3 we argued that this plane corresponds to the ecliptic.

Vaṁśīdhara goes on to suggest that SB 5.21.7 is not to be taken literally:

> SB 5.21.7 says [in paraphrase]: "The learned say the circumference of Manasottara Mountain is 95,100,000 *yojanas*." The meaning is: *mānasas* means "the moon." *Uttaraḥ* means "others beyond the moon, up to Saturn." In accordance with *jyotiṣa śāstra*, the measure of its orbit, combined with the part of Saturn, comes to 126,800,000 *yojanas*.

Here the circular Mānasottara Mountain defining the sun's orbit is interpreted indirectly to refer to the moon, Saturn, and, by implication, the planets in between. In the *jyotiṣa śāstra* the standard order of the planets is as follows: the moon, Mercury, Venus, the sun, Mars, Jupiter, and Saturn. (In Chapter 4 we have explained the relation between this and the order given in the *Bhāgavatam*.) Also, in the *Sūrya-siddhānta* the circumference of the orbit of Saturn is given as 127,668,255 *yojanas* (SS, p. 87).

Vaṁśīdhara then considers SB 5.20.43:

> According to SB 5.20.43, "The distance between the sun in the middle and the circumference of rhe universe [*aṇḍagola*, or 'egg-sphere'] is 250 million yojanas." The meaning is: "Of the sun" means "of the reflected sun." "Eggsphere" means "the circumference of the golden egg." The distance berween the two would be 250 million *yojanas*.
>
> SB 5.20.43 also says, "The sun is situared [vertically] in the middle of the universe, in the area berween *dyaus* and *bhūmi* [Bhuvarloka and Bhūrloka, or heaven and earth], which is called *antarikṣa*, outer space." The meaning of this is: *Dyāv-ābhūmyoḥ* refers to the orbit of the asterisms and the earth. *Yad antaram* means "in the middle of the egg (i.e., the golden egg)." The sun therein is the sun of the water-sphere reflection. Thus it should be understood.

This verse states that the sun is situated vertically halfway between the top and bottom of the universal egg-sphere. It lies within the region of *antarikṣa*, between the planes of Bhūrloka (or Bhū-maṇḍala) and Bhuvarloka. Vaṁśīdhara interprets the sun referred to in this verse to be a reflection of the actual sun. However, we have suggested that it can be understood as the real sun orbiting in the plane of the ecliptic. In this connection, we should note that the radius of the sun's orbit according to modern astronomy (interpreted geocentrically) is 93 million miles. For comparison, the 15,750,000-*yojana* radius of the sun's orbit around Mānasottara Mountain is about 79 million miles using 5 miles per *yojana*, and 126 million miles using 8 miles per *yojana*.

In summary, the commentary of Vaṁśīdhara on SB 5.20.38 shows that the interpretation of the Fifth Canto was a topic of doubt and controversy among Vaiṣṇavas in the 17th century. The source of doubt lay in an apparent contradiction between the Purāṇic cosmology represented by the Fifth Canto and the *jyotiṣa śāstra*. The *jyotiṣa śāstra* was regarded as the "eye of the *Vedas*," and it seemed to correspond to observable reality. Yet its description of the earth seemed totally at variance with the "big earth" of 500 million *yojanas* described in the Fifth Canto.

In this book we have argued that the contradiction between Purāṇic cosmology and *jyotiṣa śāstra* can be resolved, and that both are integral parts of an original Vedic tradition. This was also the basic point of Vaṁśīdhara's argument. He concludes,

> Thereby it is undoubtedly indisputable that by meaningful justification as to truth, the sacred *Bhāgavata*, being the statement of the supreme *āpta* [authority], is a means of proof, unrefuted, being in conformity with the conclusions of all sciences.

It follows that in Vaṁśīdhara's day, as today, an argument showing the irrefutability of the *Bhāgavatam* needed to be made. We would suggest that the material knowledge of the ancient Vedic civilization has been in disarray for a long time, and this would also be true of Vedic spiritual knowledge, were it not for Lord Caitanya and the acharyas following Him. However, this does not mean that we should denigrate the Vedic material knowledge as unrealistic and then similarly doubt the Vedic spiritual knowledge. A close examination of Vedic cosmology and astronomy suggests the presence of a deep and elaborate body of knowledge, even though today it is coming down to us in a fragmentary form.

APPENDIX 2

THE ROLE OF GREEK INFLUENCE IN INDIAN ASTRONOMY

As we pointed out in Section 1.b, Western scholars maintain that the mathematical astronomy of the *siddhāntas* was borrowed from Greek and Babylonian astronomy in the early centuries of the Christian era. In this appendix we will make a few observations suggesting that this hypothesis is not at all proven. We will not attempt an exhaustive treatment of the many arguments advanced by scholars, since this would require a large book. Rather, we will make a few points intended to show the quality of the scholars' arguments and the nature of the historical evidence used by scholars to present their case.

To begin, we should note that the history of ancient Western astronomy revolves around Claudius Ptolemy, an Alexandrian astronomer who lived in the second century A.D. Ptolemy is famous for writing a book on astronomy—the *Syntaxis*, or *Almagest*—that dominated Western astronomical thinking for over a thousand years. It turns out that, apart from Ptolemy's *Almagest*, we have very little historical evidence regarding ancient Greek astronomy. Neugebauer describes the situation as follows in his three-volume work on ancient astronomy. Concerning his discussion of Greek astronomy before Ptolemy, he says,

> With book IV [on pre-Prolemaic Greek astronomy] we entered an entirely new situation, where a later period had effaced all but vague and confused reports of its prehistory. This condition prevails right down to Ptolemy; without his historical remarks we would know almost nothing about the astronomy of Hipparchus or Apollonius (NG, p. 781).

Concerning the Roman-Byzantine period following Ptolemy, Neugebauer says,

> Over and over again attempts to see more clearly into the transmission of scientific knowledge within the Roman-Byzantine world and beyond its boundaries are made impossible by the absence of published texts. . . . It is obvious that at present any attempt at writing a historical narrative would be utterly unsatisfactory. The chances are slim that the future will be much better" (NG, p. 781).

As a result, we can divide the history of ancient Western astronomy into three periods: (1) pre-Ptolemaic, (2) the time of Ptolemy himself, and (3) post-Ptolemaic. Of these, we have substantial knowledge only of (2), as revealed by the *Almagest*. Due to our lack of solid evidence regarding period (1), we do not know the origins of Greek and Babylonian astronomy, and thus we cannot rule out the possibility that many astronomical ideas attributed to the Greeks may have come originally from India. And because of our ignorance of period (3), we have no solid basis for saying that these ideas were transmitted to India from Greek sources during the time of the decline and fall of the Roman Empire.

Nonetheless, even though our knowledge of the history of ancient astronomy is extremely incomplete, there are scholars who believe that they can uncover important parts of this history by speculative reconstruction. One example of this is a paper entitled "The Recovery of Early Greek Astronomy from India," by David Pingree (PG). In order to indicate the complexities and pitfalls of the speculative process, we will examine the key argument of this paper in detail. This will involve the use of a number of technical astronomical terms, but we will explain these as we go along. Our method will be to first present Pingree's theory, and then give his reasons for accepting this theory as true. Then step by step we will show the fallacies in his reasoning and present an alternative theory that is in better agreement with the facts.

A2.A. PINGREE'S THEORY REGARDING ĀRYABHAṬA

Pingree maintains that in the late Roman period, the Indian astronomer Āryabhaṭa used a Greek astronomical table based on Ptolemaic calculations to compute parameters for the mean motions of the planets. A planet moves at varying rates in its orbit, but one can define an artificial "average" planet that moves at a steady rate on both its primary cycle and its secondary cycle, if it has one. (Chapter 1 points out that Mercury, Venus, Mars, Jupiter, and Saturn have a secondary cycle, or epicycle.) The motion of this fictitious planet is called mean motion. To define it, two numbers are needed for each cycle: a position at a particular point in time and a rate of uniform motion. These numbers were the parameters needed by Āryabhaṭa.

Pingree proposes that Āryabhaṭa chose noon of March 21, A.D. 499, as the date for his calculations. As Pingree reconstructs it, Āryabhaṭa first used the parameters from an existing Indian astronomical text, the *Brahmapakṣa*, to compute for each planet the whole numbers of revolutions that had already elapsed from the beginning of Kali-yuga to this date.

The *Brahmapakṣa* calculations give not only the whole numbers of revolutions from the start of Kali-yuga, but also fractional parts representing the mean positions of the planets at the chosen date. According to Pingree, Āryabhaṭa knew that these mean positions were wrong. He is convinced that Āryabhaṭa was incapable of making his own observations of mean planetary positions. How then did Āryabhaṭa know that these positions were wrong? Pingree explains that a Greek astronomical table had fallen into Āryabhaṭa's hands, and he

TABLE A2.1

The Accuracy of Reconstructions of Āryabhaṭa's Parameters

Planet	R	(1)	(2)	(3)	(4)	(5)
Saturn	146,564	0	0	–12	4	0
Jupiter	364,224	4	4	–8	4	0
Mars	2,296,824	0	0	–8	4	0
Venus	7,022,388	0	–4	–16	0	0
Mercury	17,937,020	–8	–12	–24	–20	–8
Sun	4,320,000	—	4	1,192	0	0
Moon	57,753,336	8	8	–4	–36	0
Asc. Node	–232,226	0	0	–10	–88	0

This table shows the accuracy of different schemes for reconstructing Āryabhaṭa's parameters, R, for revolutions per yuga cycle of the planets. The numbered columns give the differences between the reconstructed parameters and Āryabhaṭa's actual parameters. These columns are: (1) Pingree's original results, (2) our reconstruction based on Ptolemy's mean motions relative to his position for Zeta Piscium, (3) the same, using Ptolemy's mean motions only, (4) a reconstruction obtained by rounding off the *Brahmapakṣa* periods, and (5) a reconstruction based on the hypothesis of observation.

had acquired instruction in its use from some person with Greek astronomical knowledge. On the basis of this foreign table, Āryabhaṭa knew the errors in the mean positions computed by his Indian methods, and he desired to correct them in a way that would bring glory to himself and his native India.

According to Pingree, Āryabhaṭa simply looked up the required mean positions in the Greek table. Then he converted the table's degrees, minutes, and seconds to fractions of a revolution, and added them to the whole revolutions obtained from the *Brahmapakṣa*. This gave the correct total mean motion of the planets from the start of Kali-yuga to the chosen date, assuming that the whole numbers of revolutions given by the *Brahmapakṣa* were right.

Āryabhaṭa's chosen date was exactly 3,600 of his years after the start of Kali-yuga, and he wanted to express his rates of mean motion in Indian style as numbers of revolutions in a yuga cycle, which lasts 4,320,000 years. Since 4,320,000/3,600 is 1,200, all Āryabhaṭa had to do was multiply his total mean motion figures by 1,200 and round them off to integers. (For technical reasons, Āryabhaṭa wanted these integers to be of the form 4n for the seven main planets, and 4n + 2 for Rāhu, the ascending node of the moon.)

Pingree maintains that Āryabhaṭa did this and then covered his tracks by neglecting to mention the Greek table in his astronomical writings. He also neglected to mention any of his other Greek source materials. In this way, Āryabhaṭa obtained undying fame as the author of an astronomical system of marvelous accuracy and sophistication.

A2.B. THE MAIN ARGUMENT FOR PINGREE'S THEORY

Now, how does Pingree know that this is what Āryabhaṭa did some 1,400 years ago? His key argument is that if we use Ptolemaic calculations to reproduce Āryabhaṭa's supposed steps, then we obtain Āryabhaṭa's parameters for mean planetary motion almost exactly. Āryabhaṭa's parameters, listed under R in Table A2.1, are in the hundreds of thousands and millions. Column (1) of this table lists the differences between Āryabhaṭa's parameters and these parameters as reconstructed by Pingree. For example, for Jupiter, Āryabhaṭa's rate is 364,224 revolutions per yuga cycle, and Pingree's reconstruction is larger than this by 4. Since these differences are very small, it is hard to imagine how Āryabhaṭa could have arrived at his parameters without following the scenario that Pingree proposes. This makes it seem that Pingree's conclusion concerning Āryabhaṭa is indisputable, and equally so his contention that nearly every aspect of Indian astronomy was imported from Greek sources without acknowledgement (PG, pp. 114–15).

An argument such as Pingree's has a great impact on the academic world. It tends to be immediately convincing to scholars, and it becomes established as a foundation stone in an imposing school of thought that cannot be easily challenged by nonprofessionals. As a result, scholars in other fields (such as comparative religion and history) accept the conclusions of such a school as a matter of course, and modify their own views in accordance with it.

A2.C. A PRELIMINARY CRITIQUE OF PINGREE'S ARGUMENT

However, one can indeed find other ways by which Āryabhaṭa could have arrived at his parameters. The *Brahmapakṣa* parameters are expressed in revolutions per *kalpa* of 4,320,000,000 years, whereas Āryabhaṭa wanted parameters in revolutions per yuga cycle of 4,320,000 years (see Table A2.3). What happens if we simply divide the *Brahmapakṣa* parameters by 1,000 and then round them off to suitable integers of the form 4n or 4n + 2? Column (4) of Table A2.1 shows the differences between the parameters computed in this way and Āryabhaṭa's original parameters.

We can see that for Saturn, Jupiter, Mars, and Venus the differences are not much greater than those produced by Pingree's reconstruction. For these planets we come within 4 units of Āryabhaṭa's parameters, and for Mercury, the moon, and the ascending node we come within 20, 36, and 88 units, respectively. (Pingree neglected the parameter for the sun, but we also obtain this parameter precisely.) This illustrates that Pingree's reconstruction at most accounts for the delicate fine tuning of Āryabhaṭa's parameters; most of the significant digits in these parameters come from the *Brahmapakṣa* parameters, which Āryabhaṭa acknowledges as source material.

As we shall see, this fine tuning can be accounted for in ways other than the one advocated by Pingree. To do this, it is first necessary to examine Pingree's argument more closely.

As the first step in reconstructing his calculations, we consulted Ptolemy's *Almagest* (TM2) and wrote a computer program to calculate mean planetary positions according to Ptolemy's system. In this system, mean motions are computed by linear equations, starting with initial conditions at Ptolemy's epoch of noon on February 26, 747 B.C.—the first year of the reign of King Nabonassar of Babylon. To clarify exactly what we are computing here, we will give some definitions of mean planetary positions in Indian, Ptolemaic, and modern astronomy.

In Ptolemy's system the planets Mercury, Venus, Mars, Jupiter, and Saturn move in two cycles in a way similar to the motions of these planets in the system of the *Sūrya-siddhānta* (see Chapter 1). For Mars, Jupiter, and Saturn, the mean positions in Ptolemy's system are angles measured counterclockwise on the first cycle relative to the point on the ecliptic representing the vernal equinox. In the *Sūrya-siddhanta*, the mean positions for these planets are the same, except that the reference point is the position of the star Zeta Piscium rather than the vernal equinox. The system of Āryabhaṭa is essentially the same as that of the *Sūrya-siddhanta*.

Ptolemy's system defines the mean anomalies of Mercury and Venus to be the angles measured counterclockwise on the second cycle relative to their mean positions, which are both equal to the mean position of the sun. In the *Sūrya-siddhanta* the *śīghras* of Mercury and Venus are the corresponding angles, measured with respect to Zeta Piscium. For simplicity, we will redefine the Ptolemaic mean positions of Mercury and Venus to be their mean anomalies plus the position of the sun. This agrees with Pingree's implicit usage, and provides natural quantities to compare with the *śīghras* of Mercury and Venus. We will also find it convenient to refer to these *śīghras* as the mean positions of Mercury and Venus according to the Indian system.

In the *Sūrya-siddhānta*, the position of the ascending node of the moon, or Rāhu, is defined relative to Zeta Piscium. Ptolemy's system does not directly define the motion of the moon's ascending node, but does define a related quantity called the mean motion of the moon in latitude. The position of the ascending node relative to the vernal equinox is 270° plus the difference between the moon's mean position and this quantity. In this way we can define the Ptolemaic mean position for the ascending node.

Using these definitions, we conclude that the mean positions of the planets in the Ptolemaic and Indian systems differ theoretically only in their choice of the reference point of zero longitude. In the two systems, this point is respectively the vernal equinox and the location of the star Zeta Piscium.

In modern astronomy, the mean longitudes of the planets are defined in a way that is comparable with the mean positions as we have defined them for the Indian and Ptolemaic systems. There, one measures the counterclockwise angle between the vernal equinox and the planet's heliocentric orbital position. The details can be found in texts on spherical astronomy such as SP. Here we would simply like to point out that the similarities between the Indian, Ptolemaic, and modern systems may arise as much from their describing the same planetary system as from cultural borrowing.

To find the Ptolemaic mean positions at a particular date, one determines the number of days between this date and Ptolemy's epoch and inserts this number into the equations for mean motion. For example, the traditional date

TABLE A2.2

The Ptolemaic Mean Longitudes of the Planets at Noon on March 21, A.D. 499

Planet	Ptolemy	Ptolemy Minus Zeta Piscium	Pingree
Saturn	45;56	49;19	48;40
Jupiter	185;22	188;44	188;06
Mars	4;24	7;47	7;08
Venus	351;17	354;39	356;45
Mercury	178;32	181;55	184
Sun	357;16	0;39	—
Moon	279;46	283;09	283;30
Asc. Node	−10;55	−7;32	−7;11

The rightmost column lists the Ptolemaic mean longitudes of the planets at noon of March 21, A.D. 499, as reported by Pingree in his Table 2. The leftmost column lists the Ptolemaic mean longitudes at this date, as computed by our program. The middle column lists the same figures minus the Ptolemaic position of Zeta Piscium at this date.

for the beginning of Kali-yuga is February 18, 3102 B.C. Using Āryabhaṭa's assumption that Kali-yuga began at sunrise, there are 860,172.25 days from the beginning of Kali-yuga to Ptolemy's epoch. (By convention, days begin at midnight, sunrise is .25 of a day, and noon is .5 of a day.) This figure can be used to obtain the Ptolemaic mean planetary positions at the start of Kali-yuga.

We need a way of making sure that our Ptolemaic calculations are correct. Pingree provided a way of checking this by listing the Ptolemaic mean positions of Saturn, Jupiter, Mars, the sun, the moon, and Rāhu at the Kali-yuga starting date. His figures agree precisely with ours, except in the case of Rāhu, where there is a 6-degree difference. This indicates that except for Rāhu, our program for Ptolemaic calculations agrees with Pingree's.

The star Zeta Piscium is important in Indian astronomy, since it is used as the starting point for measuring celestial longitudes along the ecliptic. We therefore wrote a program to calculate the position of this star by Ptolemaic methods, and we wanted to check the accuracy of this program.

This program is based on the following facts: According to Ptolemy's star table, Zeta Piscium had a longitude of 23° of Pisces on July 20, A.D. 137. According to Ptolemy's rule for the precession of the equinoxes, this longitude increases at one degree per century (of Egyptian 365-day years).

Pingree gave the Ptolemaic position of the star Zeta Piscium at the beginning of Kali-yuga. Calculation with our program confirms Pingree's statement that Zeta Piscium had a longitude of 320°37' at the start of Kali-yuga. We should note that in the Ptolemaic system such longitudes are measured from the vernal equinox at 0° of Aries. (These are called tropical longitudes.)

After we have checked our Ptolemaic calculations at the Kali-yuga starting date, the next step is to perform these calculations for noon of March 21, A.D. 499, the date of Āryabhaṭa's alleged calculations. There are 454,759 days from Ptolemy's epoch to this date. If we compute the Ptolemaic mean positions for this date, a number of interesting points emerge. First of all, the Ptolemaic mean longitudes do not at all agree with Pingree's figures, as given in his Table 2 (PG, p. 116). This can be seen by comparing the rightmost and leftmost columns of Table A2.2.

The middle column of Table A2.2 lists the differences between our computed Ptolemaic mean longitudes and our computed Ptolemaic position of Zeta Piscium in A.D. 499. For simplicity, we will call such differences "distances from Zeta Piscium." If we compare these figures with Pingree's reported mean longitudes in the rightmost column, we see that there is rough agreement. They differ from Pingree's reported mean longitudes by 1.2° on the average (using a root-mean-square average). This rough agreement suggests that Pingree is really listing distances from Zeta Piscium, not Ptolemaic mean longitudes. But even if this is what he intends, the agreement is still rough and should be

TABLE A2.3

A Hypothetical Reconstruction of Āryabhaṭa's Revolutions Per Yuga Cycle

Planet	Revolutions per *kalpa*	N	Ptolemy –Zeta P.	Est. 1 of R	Modern –Sun	Est. 2 of R
Saturn	146,567,298	122	49;19	146,564	48;39	146,564
Jupiter	364,226,455	303	188;44	364,228	187;29	364,224
Mars	2,296,828,522	1,914	7;47	2,296,824	7;11	2,296,824
Venus	7,022,389,492	5,851	354;39	7,022,384	356;26	7,022,388
Mercury	17,936,998,984	14,947	181;55	17,937,008	183;28	17,937,012
Sun	4,320,000,000	3,600	0;39	4,320,004	0;00	4,320,000
Moon	57,753,300,000	48,127	283;09	57,753,344	280;14	57,753,336
Asc. node	–232,311,168	–193	–187;32	–232,226	–187;43	–232,226

In this table we have reconstructed Āryabhaṭa's revolutions per *yuga* cycle (R), using revolutions per *kalpa* from the *Brahmapakṣa* and mean planetary positions according to both Ptolemy and modern calculation. The Ptolemaic mean positions are relative to the Ptolemaic position for Zeta Piscium, and the modern positions are relative to the modern position for the sun. The modern positions are computed for noon on March 21, A.D. 499, at Ujjain, and the Ptolemaic positions are computed for this date at Alexandria. The two columns of longitudes are followed by the revolutions per yuga cycle that result from them, using Pingree's method. The numbers under N are the elapsed whole revolutions, according to the *Brahmapakṣa*, from the beginning of Kali-yuga to Āryabhaṭa's 499 date.

contrasted with the precise agreement that we found for Saturn, Jupiter, Mars, the sun, and the moon at the Kali-yuga starting date.

In his Table 1, Pingree lists distances from Zeta Piscium under the heading "Distance from Zeta Piscium," and mean longitudes under lambda, the Greek letter symbolizing these quantities (PG, p. 115). Yet in his Table 2, he lists quantities under lambda that are really distances from Zeta Piscium, and he refers to these quantities as mean longitudes. We have not been able to account for this discrepancy in nomenclature.

We have also not been able to account for the discrepancies between the middle and rightmost columns of Table A2 2, for it would seem that calculations for 454,759 days after Ptolemy's epoch should be even more precise than calculations for 860,172.25 days before that epoch. (We note that Pingree's Ptolemaic calculations apparently have not been corrected for the time difference between Ptolemy's city of Alexandria and Āryabhaṭa's city of Ujjain; this possible correction does not account for the discrepancy.)

In column (3) of Table A2.1 we see the errors in reconstructing Āryabhaṭa's parameters using actual Ptolemaic mean longitudes for the selected A.D. 499 date, and not the Ptolemaic distances from Zeta Piscium used by Pingree. Clearly these errors rule out this reconstruction. In column (2) we see the errors that arise if we reconstruct Āryabhaṭa's parameters using our computed Ptolemaic distances from Zeta Piscium. These are the errors that Pingree's theory actually entails if we assume that he means distance from Zeta Piscium when he says mean longitude.

For Venus and Mercury the errors in Pingree's reconstruction of Āryabhaṭa's parameters turn out to be worse than those reported by Pingree in his paper. (Compare columns 1 and 2 of Table A2.1.) This indicates errors on Pingree's part, but it might be argued that it does not detract very badly from his hypothesis. We therefore ask, Is there some reasonable way of reconstructing Āryabhaṭa's parameters that produces smaller errors for all of the planets than Pingree's method? The answer is yes. To explain this, we must turn to a discussion of the mean positions of the planets according to modern astronomy.

A2.D. THE THEORY OF OBSERVATION

We used standard computer programs published by Duffett-Smith (DF) to calculate the mean longitudes of the planets and the moon's ascending node. We can also calculate the longitude of Zeta Piscium by looking up its position in the Astronomical Almanac and modifying this for a given date in accordance with the modern rate of 50.29 seconds per year for the precession of the equinoxes.

Let us assume, for the sake of argument, that the mean longitudes computed according to modern astronomy are correct. Then the error in Āryabhaṭa's mean position for Jupiter on a given date must be equal to Āryabhaṭa's mean position minus the position of Jupiter relative to Zeta Piscium by modern calculation. In Figure A2. 1, these errors are plotted for the eight planets for dates ranging from 10 B.C. to A. D. 1007. The vertical axis is located at noon of March 21, A.D. 499.

We can see that for the seven planets Saturn, Jupiter, Mars, Venus, the sun, the moon, and the ascending node, the errors converge sharply to a value of about 1.5° at a date near A.D. 499. (Actually, the point of closest convergence is at roughly A.D. 540.) The planet Mercury, however, is an exception to this pattern.

In Figure A2.2, similar error graphs are plotted. For these graphs we plot Āryabhaṭa's mean positions minus the corresponding differences between Ptolemy's mean positions and Ptolemy's longitude for Zeta Piscium. Here we also see a convergence at about A.D. 499. However, this convergence is much

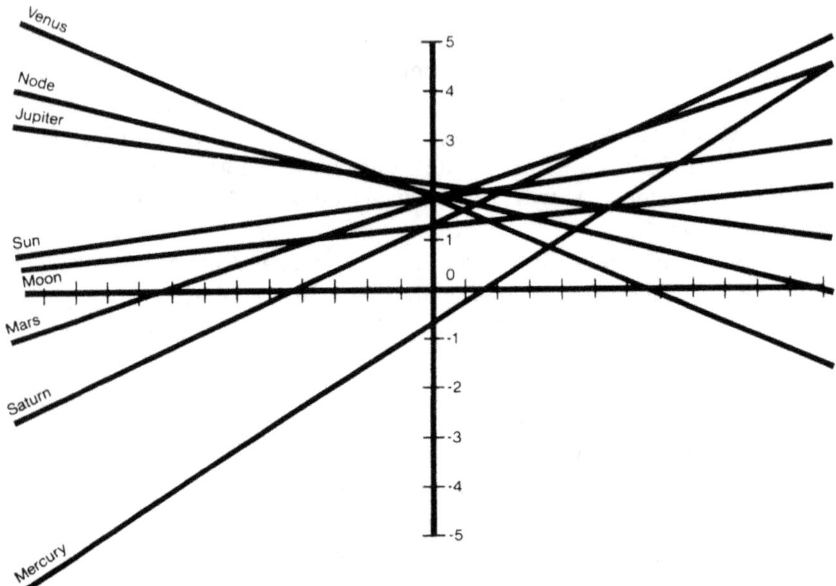

Figure A2.1 Comparison between Āryabhaṭa's system and modern astronomy. The horizontal axis represents time in unirs of 40 years. The origin corresponds to noon of Mar. 21, A.D. 499. The vertical axis represents the difference in degrees between modern mean planetary positions relative to Zeta Piscium and Āryabhaṭa's mean planetary positions. These differences are plotted for the seven planets and Rāhu (the ascending node of the moon). Note that for all planets except Mercury, the differences between Āryabhaṭa's calcularions and modern calculations converge sharply at about A.D. 539.

less sharply focused than the convergence in Figure A2.1. It is good for Saturn, Mars, the sun, and the ascending node, but it is poor for the other planets in comparison with Figure A2.1.

What is the explanation of these patterns? Pingree's argument is that the convergence in Figure A2.2 is due to the fact that Āryabhaṭa calculated his parameters so that his mean motions would agree with a Greek astronomical table at this date. But if this is so, we must ask, Why is the convergence in Figure A2.1, representing Āryabhaṭa's deviations from reality, so much sharper than the convergence in Figure A2.2, which represents his deviations from Ptolemy?

We propose the following simple answer to this question: The convergence in Figure A2.1 is due to the fact that Āryabhaṭa observed the planetary mean positions in the period between A.D. 499 and 540. The lesser convergence

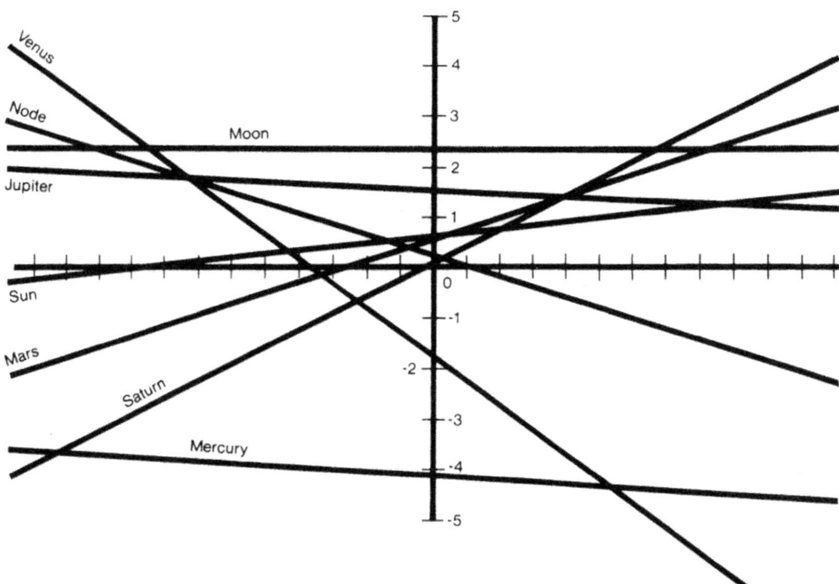

Figure A2.2 A comparison berween Āryabhaṭa's system and Ptolemaic astronomy. The horizontal axis represents time in units of 40 years. The origin corresponds to noon on Mar. 21, A. D. 499. The vertical axis represents the difference in degrees between Ptolemy's mean planetary positions relative to Zeta Piscium and Āryabhaṭa's mean planetary positions. These differences are plotted for the seven planets and Rāhu (the ascending node of the moon). In this case there is a sharp convergence only for the sun, Mars, Saturn, and the ascending node. This figure should be compared with Figure A2.1.

of plots in Figure A2.2 at this time is due to the partial agreement that exists between the Ptolemaic system and modern calculations. The convergence in A2.2 is not as sharp as that in A2.1 because there are errors in Ptolemaic mean positions relative to those computed by modern methods.

This interpretation is borne out by a comparison of Ptolemaic and modern calculations. Figure A2.3 shows plots of the difference between Ptolemaic and modern calculations of mean positions relative to Zeta Piscium. We can see that Ptolemy's errors for Saturn, Mars, the sun, and the ascending node are consistently small; the error for Jupiter is somewhat larger; and the errors for the other planets are much larger. In fact, the convergence in Figure A2.2 was strikingly good precisely for Saturn, Mars, the sun, and the ascending node.

This confirms our interpretation that the partial convergence in A2.2 is

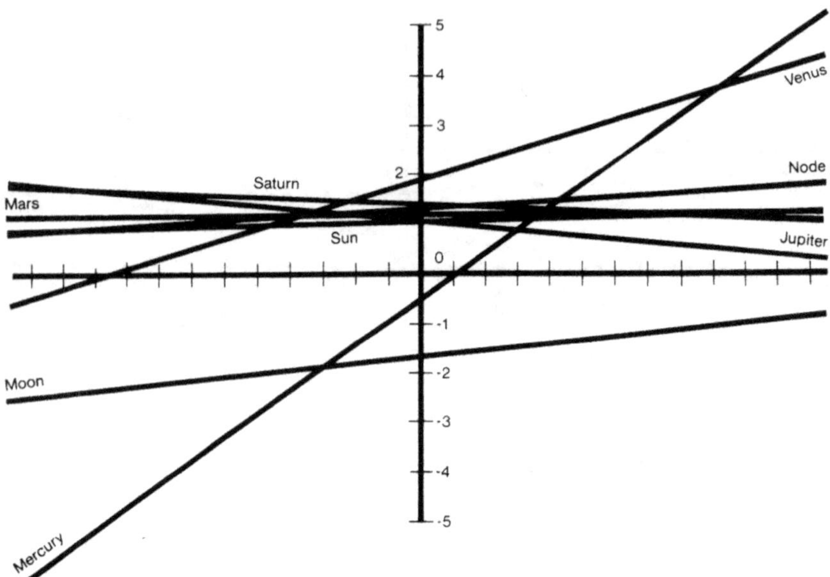

Figure A2.3 A comparison between Ptolemy's system and modern astronomy. The horizontal axis represents time in units of 40 years. The origin corresponds to Jan. 1, A.D. 161, a date in Ptolemy's lifetime. The vertical axis represents the difference in degrees between modern mean planetary positions and Ptolemy's mean planetary positions. Both the modern and the Ptolemaic mean longitudes are relative to Zeta Piscium (using modern and Ptolemaic calculations for Zeta Piscium, respectively). The differences are plotted for the seven planets and the ascending node of the moon. Ptolemy does fairly well for the sun, Mars, Saturn, Jupiter, and the ascending node, although he does make a systematic error for these planets. A much greater difference arises between Ptolemaic and modern calculations if they are both made relative to the vernal equinox. This suggests that Ptolemy's observations were initially made relative to a fixed star, and then converted to the tropical Zodiac.

simply a by-product of the greater convergence—caused by Āryabhaṭa's observations—that we see in A2.1. By Pingree's hypothesis, the scatter seen for Jupiter, Venus, and the moon in A2.2 just happens to be such that these planets converge along with the others in A2.1. This, however, seems unlikely.

At this point the argument may be raised that the convergence of error graphs in Figure A2.1 does not take place at the origin, but is about 1.5° above it. One might ask whether this can be readily explained on the hypothesis that this convergence is due to Āryabhaṭa's observations. One answer, of course, is that Āryabhaṭa may have made an error in observation that had an equal effect on all the planets. But we can go further and suggest the particular error that he may have made.

To do this we must consider the sun, which Pingree did not mention in his reconstruction of Āryabhaṭa's parameters. According to Āryabhaṭa's system, the sun is required to have a longitude of zero after 3,600 years of Kali-yuga have elapsed. (This is due to the fact that 4,320,000 is evenly divisible by 3,600.) If Āryabhaṭa found that the sun had a non-zero longitude, it would be natural for him to take this as an error and revise all his longitudes so that the longitude of the sun would come out to zero. Or, knowing that the sun should have a longitude of zero, he might have simply measured the longitudes of the other planets relative to the sun. This would automatically cause the errors in his observed longitudes to be roughly equal to the actual mean longitude of the sun at the time of his observations.

Let us suppose that Āryabhaṭa did this, and that he then computed his parameters using his observed longitudes rather than longitudes copied from a Greek table. This leads to a reconstruction of his parameters based on modern calculation of the differences between mean longitudes and the sun's mean longitude. The longitudes and resulting parameters for this reconstruction are listed in the last two columns of Table A2.3, and the errors in this reconstruction are listed in column (5) of Table A2.1. As we can see, these errors are zero, except for Mercury, where the error is equal to that in Pingree's reported reconstruction (see columns (1) and (2)). Thus, the hypothesis of observation yields better results than the hypothesis of copying from Greek tables.

A few final points will help to round out our discussion of Pingree's theory. The first is that in Figure A2.3, we can see that the Ptolemaic error graphs for several planets converge at about A.D. 161. This makes sense, since Ptolemy is thought to have written his *Almagest* at about this date. However, the convergence point is about 1.25° above the time axis. It would appear that Ptolemy too may have made some systematic observational errors.

Indeed, to properly evaluate Ptolemy's errors, we should plot the differences between Ptolemaic longitudes and modern longitudes (without making these relative to a fixed star, such as Zeta Piscium). This is because both Ptolemaic and modern longitudes are relative to the vernal equinox. If this is done, all the error curves in Figure A2.3 acquire a decided positive slope, indicating a systematic error affecting all the planets equally. (Possibly, Ptolemy's calculations were first worked out relative to a star, and then made relative to the vernal equinox using an erroneous value for the precession of the equinoxes.)

The second point is that there is no actual evidence showing that Greek astronomical tables were being transmitted to India around A.D. 500. Indeed, Neugebauer's discussion of the post-Ptolemaic period suggests that the quality of Western astronomy declined sharply after the time of Ptolemy. Thus he remarks that the astronomical material "extant from the later time of Roman Egypt is rather sad" (NG, p. 5). Of the second century work of Vettius Valens, he says, "The intervening less than 150 years succeeded not only in introducing several numerical errors into the basic parameters but also in obscuring almost completely the meaning of the prescribed operations" (NG, p. 793).

Persia is the natural link between India and the West, but of this country Neugebauer says:

> We know of Pahlavi translations of such first and second century astrological writings as Teucer and Vettius Valens and the presence of "Indian books" as well as of the "Roman *megesti*" around A.D. 250 under Shapur I. Under Khosro I . . . was revised, around A.D. 550, the famous *Zij ash-Shah*, which has been shown to be greatly dependent on Hindu sources (NG, p. 8).

Here "Roman megesti" may refer to Ptolemy, but the phrase "Indian books" suggests that Indian astronomy existed at A.D. 250 and was being exported.

Our final point is that, given the highly fragmentary nature of the surviving historical evidence, the process of speculative reconstruction is likely to create nothing more than illusions reflecting the opinions of the reconstructors. We therefore do not insist that our reconstruction of Āryabhaṭa's parameters is correct. We merely offer it as an alternative that is in better agreement with the available facts than Pingree's reconstruction.

A2.E. INDIAN TRIGONOMETRY: A SPECULATIVE RECONSTRUCTION

In the remaining part of this appendix, we will give two more examples of the process of speculative reconstruction. These examples deal with the theoretical ideas and mathematical methods of Indian astronomy, which Western historians of science say were derived entirely from Greeks or Babylonians via Greek intermediaries.

Our first example concerns the trigonometry used in texts of Indian mathematical astronomy. Our modern trigonometry is usually traced back to the Arabs (PF, p. 260). However, in the *Sūrya-siddhānta*, as well as in texts by Āryabhaṭa and other Indian astronomers, sines and cosines are used, and a table of sines is given. A modern sine is defined geometrically using a unit circle, and the corresponding Indian sine is defined in the same way, using a circle with a radius of 3,438. This means that each sine is 3,438 times as large as its modern counterpart. It also means that if angles are expressed in minutes of arc, then the sine of a small angle is nearly equal to that angle. This useful feature is achieved in modern mathematics by measuring angles in radians, a technique first invented in England in 1783 (PF, p. 270).

Another feature of the number 3,438 is that it represents a close approximation to pi. If the circumference of a circle is divided into 21,600' (i.e., 360° times 60 minutes/degree), then the circumference divided by 2π is 3,437.746, or 3,438 to the nearest integer. Thus if one wishes to work with whole numbers, 3,438 is the best value for the radius of a circle of this circumference.

Here is what some prominent historians of science have to say about the Indian sine tables:

(1) Neugebauer: "The decisive step in proving that the Indian table of sines was derived from the Hipparchian table of chords was made by G. J. Toomer" (NG, p. 299).

(2) B. L. van der Waerden: "C. G. [sic] Toomer has shown that the chord table of Hipparchus was a table of chords in a circle of radius R=3,438. . . . Toomer is justified in concluding that Āryabhaṭa's table of sines was derived from Hipparchus' table of chords by halving the chords" (VW, p. 211).

(3) D. Pingree: "This Indian sine-table is closely related to Hipparchus' chord-table as reconstructed by Toomer, in which R also is 3,438" (PG, p. 114).

These statements certainly convey the impression that the Indian sine table was directly obtained from a related trigonometrical table used by the Greek astronomer Hipparchus. However, what do we find if we actually examine the paper by G. J. Toomer that these authorities are citing? Let us briefly consider this.

The first thing that we learn from this paper is that there are no surviving Greek documents containing Hipparchus' chord table, even in a fragmentary form. Indeed, "there is no explicit evidence about the nature of Hipparchus' chord table," and no real proof that such a table ever existed (TM1, p. 6). It is important to note that only one work of Hipparchus' has survived—a commentary on the stars—and this does not present his mathematical methods. As we have already noted, this is typical of the state of our knowledge of pre-Ptolemaic Greek astronomy.

(The chord of an angle is defined as follows: Extend the sides of the angle until they intersect a circle of unit radius centered on the angle. The chord of the angle is defined to be the length of the chord of the circle connecting the two points of intersection. The chord of an angle is therefore twice the sine of half the angle.)

Having admitted that he has no direct evidence regarding his hypothetical chord table, Toomer proceeds to construct the table from scratch. He does this using methods taken directly from works of Indian astronomy. Since in these works the sine of an angle is 3,438 times the corresponding modern sine, Toomer creates a chord table in which the chords are 3,438 times the corresponding modern chords. (These are computed using a modern sine table.) He also tabulates his chords at intervals of 7.5° or twice the interval of 3.75° typically used in Indian sine tables.

To justify his construction, Toomer uses it to show how Hipparchus might have arrived at two numbers describing the moon's orbit that are ascribed to him by Ptolemy. Since we do not actually know what computational methods Hipparchus used, Toomer takes it for granted that he used certain methods of Ptolemy. Using these methods, plus his hypothetical chord table, Toomer computes one of Hipparchus' numbers, but gets it wrong. He then argues that Hipparchus must have made a particular mistake in the complex procedure. When he computes the number again on this basis, it still comes out wrong (3,082⅔ over 246⅓, rather than 3,122½ over 247⅔). But Toomer concludes that it is close enough to "prove" that Hipparchus did use a chord table of the proposed type, and that he made the proposed mistake (TM1, p.12). The second number also comes out wrong (3,134 over 338 rather than 3,144 over 327⅔), but Toomer again regards it as close enough.

By this reasoning Toomer maintains that "the nature of Hipparchus' chord table is conclusively established" (TM1, p.16). Since the table has the structure of an Indian sine table, it follows that Indian trigonometry must have been derived from the Greeks. The idea that Greeks may have been influenced by Indian developments is never even suggested by modern Western historians of science. But in this case, of course, we have no evidence for influence either way, since the connection between Hipparchus' two numbers and the Indian sine table is purely speculative.

Besides his interpretation of two numbers in the *Almagest*, Toomer offers only one other piece of evidence suggesting that the Greeks used a chord table with a radius of 3,438. This is a statement in Ptolemy's Geography mentioning for two cases the ratio between the length of a parallel of latitude and the length of the equator. For Rhodes, at 36° north, this ratio is 93/115, and for Thule, at 63°, it is 52/115. Toomer claims that these figures must have been derived from Hipparchus' hypothetical chord table, since in that table the diameter, expressed in degrees, rounds to 115. Also the chords corresponding to the two latitudes turn out to be 93 and 52 when read from that table by linear interpolation and converted from minutes to degrees. According to Toomer, "The conclusion seems inevitable that he [Ptolemy] is here using, directly or indirectly, the old chord table of Hipparchus" (TM1, p. 25).

Yet there are many ways in which Ptolemy might have arrived at these numbers. For example, he might have reasoned that it would be useful to use a degree of latitude as a unit of distance in geographical studies. (In fact, a degree of latitude was sometimes assumed by the ancient Greeks to have a length of 700 stades, or about 80 miles (NT, p.45).) In this case the diameter of the earth would be the circumference of 360 units, divided by π. Using Archimedes' rough estimate of 22⁄7 for π, this diameter is 115 to the nearest unit. Using a compass, a ruler, and a protractor, it is easy to construct a circle of this diameter, marked with the parallels of latitude at 36° and 63°. Their lengths turn out to be 93 and 52 in round numbers.

In fact, we performed this construction in about 10 minutes; it is much easier to obtain the ratios in this way than by using linear interpolation in a table of chords. Thus there is no need to suppose that a chord table, with or without a radius of 3,438, was ever involved.

A2.F. ANOTHER SPECULATIVE RECONSTRUCTION

Since the chord table of Hipparchus has not survived (if it ever existed), it is remarkable that such slender evidence can be offered as the basis for "inevitable" conclusions about it. Yet, as we have seen, such speculative reconstructions are not unusual in the field of the history of science. Here we will give one more example. This is provided by the mathematician B. L. van der Waerden, who traces back Hipparchus' trigonometry to the Greek mathematician Apollonius of Perge (VW, pp. 211–12). Van der Waerden's reasoning goes as follows:

(1) The Indian sine tablets accompanied by a complete theory of trigonometry, as shown by the writings of Āryabhaṭa. This too must have come from the Greeks, but Hipparchus, in van der Waerden's estimation, was not a good enough mathematician to have invented it.

(2) This mathematician could not have been Archimedes, since he used 22⁄7 for π. Therefore it must have been an able Greek mathematician living between the times of Archimedes and Hipparchus.

(3) There was exactly one excellent mathematician living in this period, namely Apollonius of Perge. Now, Eutokios, in a commentary on Archimedes, says that Archimedes' estimate of pi was intended for "the needs of daily life," and that Apollonius had given more accurate estimates.

(4) On this basis, "we are bound to conclude" that the value of π used in Indian trigonometry is due to Apollonius (VW, p. 212).

(5) In fact, the Indian astronomer Bhāskarācārya gives 3927/1250 (3.1416) as a good estimate of π, and also gives 22⁄7 as an estimate "adopted to practice." Since this statement is very similar to Eutokios' statement about Archimedes and Apollonius, "we are bound to conclude that they go back to a common source, and hence that the estimate of π is due to Apollonius" (VW, p. 212).

One should note here that van der Waerden does not cite a reference giving Apollonius' estimate for π, and he also gives no reference that specifically attributes studies of trigonometry to Apollonius. Thus we do not know what Apollonius' estimate of π was, nor do we know whether he actually knew any trigonometry. Nor do we know whether Apollonius was the only able mathematician living between Hipparchus and Archimedes. And even if he was, we do not know who invented the basic theory of trigonometry, when this was done, or in what country that person lived. Van der Waerden's argument is simply a chain of suppositions.

We have discussed the arguments of Pingree, Toomer, and van der Waerden in detail to show the kind of foundations that underlie scholarly conclusions about the origins of Indian astronomy. The main characteristic of these foundations is that they are composed almost entirely of unsupported assumptions, biased interpretations, and imaginary reconstructions. It is unfortunate, however, that after many scholars have presented arguments of this type in learned treatises, the arguments accumulate to produce an imposing stratified deposit of apparently indisputable authority. In this way, supposedly solid facts are established by the fossilization of fanciful speculations whose original direction was determined by scholarly prejudice. Ultimately, these facts are presented in elementary texts and popular books, and accepted on faith by innocent people.

The arguments of Toomer and van der Waerden are clearly very weak. But the objection might be raised that the division of the circle into 21,600' in Indian trigonometry is itself evidence of Greek influence. In answer to this, we should first point out that according to modern scholars, the division of the circle into degrees, minutes, and seconds was borrowed by the Greeks from the Babylonians. We therefore ask, Did the Babylonians invent this division, or might they have borrowed it from some other source?

In fact, there is evidence that the division of a circle into 360° is very old, and is related to the number of days in a year. In the *Śrīmad-Bhāgavatam* the number of days in a year is given repeatedly as 360 (see SB 3.11.10–12, for example). The same number is given in the *Ṛg Veda*, which is accepted even by Western scholars as dating back to 1000–1200 B.C. (HA, p. 8). For example, in the *Ṛk-saṁhitā*, it is stated,

> Twelve spoke-boards, one wheel, three navels. Who understands these? In these there are 360 Sankus (rods) put in like pegs which do not get loosened (BJS, p. 18).

This verse speaks of a year as having 360 days, and it can be compared with a similar statement in SB 5.21.13, in which the year is also described as a wheel. There are many statements in the Vedic literature comparing the year to a wheel or circle.

The 360-day year was kept in alignment with the seasons by periodically inserting an intercalary month. This is described in Śrīla Prabhupāda's purport to SB 5.22.7.

The time accepted by scholars for the *Ṛg Veda* antedates the known period of Babylonian astronomy. According to Neugebauer, Babylonian astronomy dates back no further than about 600 B.C.:

> We know very little about the prehistory of this Babylonian astronomy. In the extant texts from the Hellenistic period almost all methods appear fully developed. On the other hand it is virtually certain that they did not exist at the end of the Assyrian period. Thus one must assume a rather rapid development during the fourth or fifth century B.C. (NG, pp. 3–4).

We would suggest that the division of the circle into 360° was an ancient feature of Vedic civilization. In Egypt and Mesopotamia it may also date back to times when the civilizations of the Near East were part of a larger Vedic world system. As far as we are aware, this is neither demonstrated nor contradicted by known historical evidence. As the above statement by Neugebauer indicates, we have practically no historical evidence regarding the early history of astronomy in the Near East.

If the division of the circle into degrees corresponds to the 360-day year, then its division into 12 signs of the zodiac, each with 30°, may correspond to the 12 Vedic months of 30 days. Likewise the Greek *bathmoi*, or 15-degree intervals, may correspond to the 15-day bright and dark fortnights of the moon. Going further, we note that among the many Indian time divisions there is the *ghaṭikā*, which is one sixtieth of a day. Also, the *pala* is one sixtieth of a *ghaṭikā*, and the *vipala* is one sixtieth of a *pala*. Next comes the *prativipala*, which is one sixtieth of a *vipala* (BJS, part 2, p. 13). Do these correspond to the divisions of a degree into minutes, seconds, and so on? We can only speculate about the ultimate origins of such divisions.

As a final point, we should note that the assumption of the Western historians of science seems to be that no one in India could have exhibited mathematical or scientific inventiveness, and thus all Indian mathematical astronomy must have been due to Western creativity. However, the available historical evidence seems to contradict this. For example, the 14th-century Indian mathematician Mādhava gave the following approximation for π:

$$\frac{2{,}827{,}433{,}388{,}233}{900{,}000{,}000{,}000} = 3.14159265359$$

Mādhava showed great creativity in his mathematical work, and is credited with inventing the power series expansion for the arc-tangent function, which was separately discovered in Europe by James Gregory in 1671 (SA, p. 182). Since Mādhava lived in a traditional Indian cultural milieu, such mathematical creativity has presumably been available in India for thousands of years.

BIBLIOGRAPHY

I.

Works by His Divine Grace A. C. Bhaktivedanta Swami Prabhupāda.These works are all published by the Bhaktivedanta Book Trust in Los Angeles, California.

BG: *Bhagavad-gītā As It Is* (1983)

CC: *Śrī Caitanya-caritāmṛta* (1974)

CN: *Conversations with Śrīla Prabhupāda*, vol. 1. (1988)

KB: *Kṛṣṇa, the Supreme Personality of Godhead* (1-vol. edition, 1986)

LB: *Light of the Bhāgawata* (1984)

NOD: *Nectar of Devotion* (1982)

SB: *Śrīmad-Bhāgavatam* (1987)

TLC: *Teachings of Lord Caitanya* (1985)

TQK: *Teachings of Queen Kuntī* (1978)

II.

Other works.

AA: Clark, Walter E., trans., *The Aryabhatiya of Aryabhata* (Chicago: Univ. of Chicago Press, 1930).

AAA: "Air Force Observations of an Unidentified Object in the South-Central U.S., July 17, 1957," *Astronautics and Aeronautics*, July 1971, pp. 66–70.

ABW: Agarwala, G. C., ed., *Age of Bhārata War* (Delhi: Motilal Banarsidāsas, 1979).

AL: Sachau, Edward C., trans., *Alberuni's India* (London: Kegan Paul, Trench, Trubner & Co., 1910).

AR 1: Arp, H., "Observational Paradoxes in Extragalactic Astronomy," *Science*, vol. 174 (1971), pp. 1189–1200.

AR2: Arp, H., and J. W. Sulentic, "Analysis of Groups of Galaxies with Accurate Redshifts," *Astrophysical Journal*, vol. 291 (1971), pp. 88–111.

AR3: Arp, H., "Distribution of Quasistellar Radio Sources on the Sky," *The Astronomical Journal*, vol. 75 (1970), no. 1, pp. 1–12.

AR4: Arp, H., "NGC-1199," *Astronomy*, vol. 6 (1978), p. 15.

AS: Kulkarni, S. D., ea., *Ādi Śaṅkara* (Bombay: Shri Bhagavan Vedavyasa Itihasa Samshodhana Mandira, 1987).

BB: Śrīla Sanātana Goswāmī, *Sri Brihat Bhāgavatāmritam* (Madras: Sree Gaudiya Math, 1975).

BD: Chambers, R., *The Book of Days* (Detroit: Gale Research Co., 1967).

BJS: Dikshit, Sankar Balakrishna, *English Translation of Bharatiya Jyotish Śāstra* (Calcutta: Gov. of India Press, 1969).

BR1: Burbidge, G., "The Line-Locking Hypothesis . . . ," *Physica Scripta*, vol. 17 (1978), pp. 237–41.

BR2: Burbidge, G., "Evidence for Non-cosmological Redshifts," *International Astronomical Union Symposium No. 92: Objects of High Redshift*, G. O. Abell, ed. (Boston: Reidel Co., 1980), pp. 99–105.

BS1: Śrīmad Bhakti Pradip Tirtha, *Śrīla Sarasvatī Thakur*, 2nd ed. (Calcutta: Gaudiya Mission, 1978).

BS2: Tridandisvami Bhaktikusum Sraman, *Prabhupāda Śrīla Sarasvatī Ṭhākura* (Sree Mayapur: Sri Chaitanya Math).

CH: Cantor, G. N. and M. J. Hodge, eds., *Conceptions of Ether, Studies in the History of Ether Theories: 1740–1900* (Cambridge: Cambridge Univ. Press, 1981).

CR1: Corliss, W. R., *Mysterious Universe* (Glen Arm, Md.: The Sourcebook Project, 1979).

CR2: Corliss, W. R., *The Moon and the Planets* (Glen Arm, Md.: The Sourcebook Project, 1985).

DF: Duffett-Smith, P., *Astronomy with Your Personal Computer* (Cambridge: Cambridge Univ. Press, 1985).

DR: Rawlins, D., "The Mysterious Case of the Planet Pluto," *Sky and Telescope* (March 1968), pp. 160–62.

DS: Waters, T., "Gravity Under Siege," *Discover* (April 1989), pp. 18–20.

EA: Motz, L., and A. Duveen, *Essentials of Astronomy* (New York: Columbia Univ. Press, 1977).

EB: Brown, J. E., ea., *The Sacred Pipe* (Baltimore: Penguin Books, 1971).

ET: Andrews, G. C., *Extra-Terrestrials Among Us* (St. Paul, Minn.: Llewellyn Pub., 1987).

EW: Evans-Wentz, W. Y., *The Fairy-Faith in Celtic Countries* (New York: University Books, 1966).

FJ: Johnston, F., *Fatima* (Rockford, Ill.: Tan Books, 1979).

GP: Sarma, K.V., trans., *The Goladīpikā by Parameśvara* (Madras: The Adyar Library and Research Centre, 1956).

GS: Simpson, G. G., *This View of Life* (New York: Harcourt, Brace, and World, 1964).

HA: Kay, G. R., *Hindu Astronomy* (New Delhi: Cosmo Pub., 1981).

HM: de Santillana, G. and H. von Dechend, *Hamlet's Mill* (Boston: Gambit, 1969).

JV: Vallee, J., *Dimensions* (Chicago: Contemporary Books, 1988).

JV2: Vallee, J., *Passport to Magonia* (Chicago: Henry Regnery Co., 1969).

MN: Thompson, R., *Mechanistic and Nonmechanistic Science* (Los Angeles: Bhaktivedanta Book Trust, 1981).

MSF: Baker, D., *The History of Manned Space Flight* (New York: Crown, 1981).

ND: Needham, J., *Science and Civilization in Ancient China*, vol. 3 (Cambridge: Cambridge Univ. Press, 1959).

NG: Neugebauer, O., *A History of Ancient Mathematical Astronomy* (Berlin: Springer-Verlag, 1975).

NM: Bhaktivinoda Ṭhākura, *Navadwip Mahatmya*, trans. Banu dāsa, ms.

NT: Newton, R. R., *The Crime of Claudius Ptolemy* (Baltimore: Johns Hopkins Univ. Press, 1977).

PA: Phillimore, J. S., *Philostratus in Honour of Apollonius of Tyana* (Oxford: Clarenden Press, 1912).

PF: Playfair, J., trans., *On the Trigonometry of the Brahmins* (Royal Soc. of Edinburgh, 1798).

PG: Pingree, D., "The Recovery of Early Greek Astronomy from India," *Journal for the History of Astronomy*, vol. vii (1976), pp. 109–123.

PL: Lowell, P., *The Evolution of Worlds* (New York: Macmillan, 1909).

PN: Hartney, W. "The Pseudoplanetary Nodes of the Moon's Orbit in Hindu and Islamic Iconographies," *Ars Islamica*, vol. 5 (1938).

PR: Procter, R., *Old and New Astronomy* (London: Longmans, Green, and Co., 1892).

RC: Field, G. B., H. Arp, J. N. Bahcall, *The Redshift Controversy* (Reading, Mass.: W. A. Benjamin, Inc., 1973).

RP: Jaki, S., *The Relevance of Physics* (Chicago: The Univ. of Chicago Press, 1970).

RS: Story, R., *Sightings* (New York: Quill, 1982).

RS2: Story, R., *Guardians of the Universe?* (New York: St. Martins Press, 1980).

SA: Sarasvatī Amma, T. A., *Geometry in Ancient and Medieval India* (Delhi: Motilal Banarsidāsas, 1979).

SBS: Bhaktisiddhānta Sarasvatī Goswāmī Ṭhākura, *Śrī Brahma-saṁhitā* (Los Angeles: Bhaktivedanta Book Trust, 1985).

SH: Eliade, M., *Shaminism* (Princeton: Princeton Univ. Press, 1964).

SK: Silk, J., *The Big Bang* (San Francisco: W. H. Freeman, 1980).

SP: Green, R. M., *Spherical Astronomy* (Cambridge: Cambridge Univ. Press, 1985).

SS: Sastrin, Bapu Deva, trans., *Sūrya-siddhānta* (Calcutta: Baptist Mission Press, 1860, reprinted in *Bibliotheca Indica*, New Series No. 1, Hindu Astronomy I).

SSB1: Wilkinson, L., trans., *Siddhānta-śiromaṇi of Bhāskarācārya*, rev. by B. D. Sastrin (Calcutta: Baptist Mission Press, 1861, reprinted in *Bibliotheca Indica*, New Series No. 1, Hindu Astronomy I).

SSB2: Arkasomayaji, D., trans., *Siddhānta Siromani of Bhāskarācārya* (Kendriya Sanskrit Vidyapeetha, Tirupati Series No. 29, 1980).

SU: Sulentic, J., "Confirmation of the Luminous Connection Between NGC 4319 and Markarian 205," *Astrophysical Journal*, vol. 265 (1983), pp. L49–L53.

SW: Swerdlow, N. M., *Ptolemy's Theory of the Distances and Sizes of the Planets* (Ann Arbor, Michigan: University Microfilms, Inc., 1969).

TD: Dobzhansky, T., "Darwinian Evolution and the Problem of Extraterrestrial Life," *Perspect. Biol. Med.*, vol. 15 (1972), pp. 157–75.

TF1: Tifft, W. G., "Periodicity in the Redshift Intervals for Double Galaxies," *Astrophysical Journal*, vol. 236 (1980), pp. 70–74.

TF2: Tifft, W. G., "Discrete States of Redshift and Galaxy Dynamics II . . . ," *Astrophysical Journal*, vol. 211 (1977), pp. 31–46.

TF3: Tifft, W. G., and W. J. Cocke, "Global Redshift Quantization," *Astrophysical Journal*, vol. 287 (1984), pp. 492–502.

TF4: Tifft, W. G., "Absolute Solar Motion and the Discrete Redshift," *Astrophysical Journal*, vol. 221 (1978), pp. 756–75.

TF5: Tifft, W. G., "Discrete States of Redshift and Galaxy Dynamics III . . . ," *Astrophysical Journal*, vol. 211 (1977), pp. 377–91.

TF6: Tifft, W. G., "Quantum Effects in the Redshift Intervals for Double Galaxies," *Astrophysical Journal*, vol. 257 (1982), pp. 442–49.

TF7: Cocke, W. J., and W. G. Tifft, "Redshift Quantization in Compact Groups of Galaxies," *Astrophysical Journal*, vol. 268 (1983), pp. 56–59.

TM1: Toomer, G. J., "The Chord Table of Hipparchus and the Early History of Greek Trigonometry," *Centaurus*, vol. 18 (1973–74), pp. 6–28.

TM2: Toomer, G. J., *Ptolemy's Almagest* (London: Duckworth, 1984).

TSA: Smart, W. M., *Textbook on Spherical Astronomy* (Cambridge: Cambridge Univ. Press, 1962).

VG1: Vigier, J. "Cosmological Implications of Non-velocity Redshifts—A Tired Light Mechanism," *Cosmology, History, and Theology*, W. Yourgrau, ed. (New York: Plenum Press, 1977), pp. 141–57.

VG2: Jaakkola, T., M. Moles, J. P. Vigier, J. C. Pecker, and W. Yourgrau, "Cosmological Implications of Anomalous Redshifts . . . ," *Foundations of Physics*, vol. 5 (1975), no. 2, pp. 257–69.

VG3: Kuhi, L., J. Pecker, and J. Vigier, "Anomalous Redshifts in Binary Stars," *Astronomy and Astrophysics*, vol. 32 (1974), pp. 111–14.

VG4: Merat, P., J. Pecker, and J. Vigier, "Possible Interpretation of an Anomalous Redshift . . . ," *Astronomy and Astrophysics*, vol. 30 (1974), pp. 167–74.

VJ: Shamasastry, R., *Vedangajyautisha* (Mysore: Asst. Supt., Govt. Branch Press, 1936).

VNB: Thakur Bhaktivinod and Siddhānta Saraswati, *Vaishnavism and Nambhajan* (Madras: Sri Gaudiya Math, 1968).

VP: Wilson, H. H., trans., *The Vishnu Purāṇa*, vol. 1 (London: Trubner & Co., 1864).

VR1: Varshni, Y. P., "Alternative Explanation for the Spectral Lines Observed in Quasars," *Astrophysics and Space Science*, vol. 37 (1975), L1–L6.

VR2: Varshni, Y. P., "The Redshift Hypothesis for Quasars: Is the Earth the Center of the Universe?" *Astrophysics and Space Science*, vol. 43 (1976), pp. 3–8.

VR3: Varshni, Y. P., "The Redshift Hypothesis for Quasars: Is the Earth the Center of the Universe? II." *Astrophysics and Space Science*, vol. 51 (1977), pp. 121–24.

VSB: Vasu, Rai Bahadur Srisa Chandra, trans., *The Vedanta-sutras of Badarayana* (New Delhi: Oriental Books Reprint Corp., 1979).

VW: van der Waerden, B. L., *Geometry and Algebra in Ancient Civilizations* (Berlin: Springer-Verlag, 1983).

WG: Wigner, E., "Physics and the Explanation of Life," *Foundations of Physics*, vol. 1 (1970), no. 1, pp. 35–45.

TABLES

ILLUSTRATIONS

INDEX

www.ingramcontent.com/pod-product-compliance
Lightning Source LLC
La Vergne TN
LVHW020709110826
845149LV00012B/2182

* 9 7 8 0 9 9 8 1 8 7 1 5 0 *